99 WAYS
TO A SIMPI
LIFESTYLE

The Center for Science in the Public Interest is a non-profit research organization in Washington, D.C., that investigates public-interest issues, including energy, environmental protection, consumer safety, and health nutrition. Through its activities and with the help of public participation, CSPI keeps a watch over business interests and corporate power's tendency to disregard the welfare of the public and the environment.

Albert J. Fritsch, S.J., Ph.D., is an organic chemist and co-director of the Center for Science in the Public Interest. He has been particularly involved in areas concerning environmental protection and resource conservation and appears frequently before congressional committees on these and other consumer issues.

99 WAYS TO A SIMPLE LIFESTYLE

by the Center for Science in the Public Interest

SIMPLE LIFESTYLE TEAM

Albert J. Fritsch (Director)
Barbara O. Hogan
Dennis J. Darcey
David E. Taylor
A. S. Csaky (Illustrator)

ANCHOR BOOKS
ANCHOR PRESS/DOUBLEDAY
GARDEN CITY, NEW YORK
1977

99 WAYS TO A SIMPLE LIFESTYLE was originally published by the Center for Science in the Public Interest.

Library of Congress Cataloging in Publication Data
Main entry under title:

99 ways to a simple lifestyle.

Includes bibliographical references and index.
1. Home economics. 2. Energy conservation.
3. Consumer education. I. Center for Science in the Public Interest.
TX147.N7 1977 640

Anchor Books Edition: 1977

ISBN 0-385-12493-7
Library of Congress Catalog Card Number 76-54756

Printed in the United States of America

ACKNOWLEDGMENTS

We would like to thank the many dedicated people, both staff and volunteers of CSPI, who have helped assemble the "99 ways." Among those who deserve special mention are the following: Brian Owens (community), Mary Anne Soccio (clothing), Thomas Green (instructional diagrams), Dorothy Moursund (baking bread), Br. Thomas Conry (personal products), Cheryl Tennille (toys and relaxation), Marc Worthington and Susan Schiffrin (research), Chuck Smith and Sandy Adams (preserving food and general lifestyle philosophy), Mary Lou Doepker (cosmetics), JoAnn Scott (conservation), and Dr. Arthur Purcell (solid waste). We would like to thank the following for constructive criticism: Drs. William Millerd, Michael Jacobson and James Sullivan; Ken Bossong, Debbie Purcell, and Patti Hausman. This could not have been assembled without the technical assistance of Jackie Sabath, Shirley Larson, and Carol Leath, and the management skills of Jane Robinson, Linda Wichmann, and Arlene Semple. We are also indebted to Alan Okagaki and David Fitz-Patrick for assisting with introductory materials.

I do not think that any civilization can be called complete until it has progressed from sophistication to unsophistication, and made a conscious return to simplicity of thinking and living.

Lin Yutang
The Importance of Living, 1938

CONTENTS

INTRODUCTION

The environmental crisis that we face today is hardly news to anyone. Americans are bombarded daily with reports of fuel shortages, polluted air and water, chemically contaminated food, diminishing resources and ecologically destructive technology. Worldwide hunger and spiraling population growth are even more critical components of the environmental dilemma. Most disturbing of all is the role of Americans who, while comprising 5.4 per cent of the world's population, use 30 per cent of the world's natural resources.

We must realize that we can no longer continue our wasteful, overconsumptive habits in a world of finite resources. With this realization comes a movement toward a new way of life, or lifestyle, compatible with nature and supportive of peoples around the world.

A December 4, 1975, Harris poll indicates that 77 per cent of Americans are willing to simplify their lifestyles. With this willingness in mind, this book offers 99 practical suggestions to simplify our lifestyles while still enjoying healthy, rewarding lives. These hints are meant to improve our quality of life by reducing resource demand and pollution, and by encouraging us to devote more time to social action and political change.

Reasons for choosing a simple lifestyle might include some of the following:

- Naturalistic–helps us appreciate the serenity of nature, its silence, the changes of season, and its creatures.
- Symbolic–promotes solidarity with the world's poor, and reduces the hypocrisy of our current overconsumptive lifestyle.

- Person-oriented—affords greater opportunities to work together and share resources with one's neighbors.
- Ecological—reduces our use of resources, lessens pollution, and creates an awareness that we must live in harmony with our world.
- Health—lessens tension and anxiety, encourages more rest and relaxation, reduces use of harmful chemicals, creates inner harmony.
- Economic—saves money, reduces the need to work long hours, and increases both number and quality of jobs.
- Spiritual—allows time for meditation and prayer, and rejects materialistic values.
- Social—induces frustration with the limited scope of individual action and incites us to move to social and political action levels.

To a certain extent, the nature of simple lifestyle will be dictated by environmental constraints, but it will also be a product of our imagination and ingenuity. Moving toward a simpler lifestyle involves both qualitative and quantitative change. We must remake the way we live; the whole array of cultural, social, physical, and psychological factors that influence our mode of living must be critically examined. But what exactly is "simple lifestyle"?

"Simple lifestyle" has a variety of meanings, but is limited in this book to ways of conserving material and human resources, be these food, fossil fuel, water, wildlife and community, or individual physical and psychological health. We should not be directly concerned with a simple lifestyle that is isolated or unadorned. Our goal cannot be a return to the backwoods, for such a lifestyle is beyond the realm of possibility for many Americans. The following hints are good examples of the broad definition of simple lifestyle:

Type of Conservation	*Sample Practice in "99 Ways"*
Food	#25—Consume Less Meat
Fossil Fuel	#79—Encourage Mass Transit
Materials	#48—Ban the Non-returnable
Water	#16—Save Water

Community Health	#97–Fight Environmental Pollution
Individual Health	#68–Do Not Abuse Drugs
Wildlife	#55–Do Not Wear Fur from Endangered Species

When examining the definition of "simple lifestyle" a number of qualities surface. For example, simple lifestyle practices are less wasteful, less showy and fashion-oriented, capable of reusing items, less consumptive and not addicted to commercialism, not overly mobile, not noisy. They are more natural than synthetic, homemade than factory produced, more personal than institutionalized, with more human involvement than energy use.

Quality	*Sample Practice in "99 Ways"*
Reusable	#54–Mend and Reuse Garments
Resourcefulness	# 1–Save Heat and Insulate
Not Superfluous	#24–Eliminate Unnecessary Appliances
Do-it-yourself	#45–Keep Bees
Stability	#81–Avoid Unnecessary Auto Travel
Homemade	#34–Bake Bread
Peace and Quiet	#61–Preserve a Place for Quiet
Natural	#29–Eat Wild Foods
Personal	#90–Care for the Elderly and Ill
Human Effort	#64–Exercise Without Equipment

By striving to live a simpler lifestyle we open ourselves to a greater variety of new experiences and values. We create a non-competitive atmosphere where people can reduce the tensions of modern life and curb excesses that shorten that life. If we live at a slower pace, we have time to see the simple joys, especially the natural wonders around us. New emphasis is placed on forgotten arts and crafts. We begin to "need" far less than we previously thought.

The argument most often heard in favor of our current practices is convenience; they save human energy and allow more leisure time. Some current practices do save time, but for what reason? So we can jump out into a traffic jam heading to the nearest beach or concert? It takes longer to drive across an urban area today than it did a century ago. An electric gadget may save time in the home, but there are many benefits in allotting time for preparing meals rather than paying to keep the gadget running. As Ivan Illich observes, the typical American devotes more than 1,600 hours a year to his car. We sit in it, search for parking places and work to pay for it with 28 per cent of our income. Is this really convenience?

Another misleading argument is that the current American lifestyle is better because the rest of the world is striving to imitate it. While examples of attempts to imitate our lifestyle are apparent in various countries, it is chauvinistic to think that is due to good quality. Concerned citizens, especially in the Third World abhor the move toward expenditure of precious resources on frivolous consumer goods, and regard the trend toward excessive consumerism as the deliberate result of American commercial pressure to buy more goods.

The third argument for continuing current American lifestyles is national economic health. Consumer demand is thought to create more jobs. A closer look at current practices suggests the opposite. For example, non-returnables have cut the need for many jobs in smaller soft-drink and beer establishments. Mass production methods have destroyed numerous handcraft and small businesses that supplied employment in former times. Industry emphasizes weapons and mass-produced throwaway items that are energy, not labor, intensive. A shift toward simpler lifestyles will lead to a healthier economy where more people are employed. In parts of the economy where some job displacement will inevitably occur, service jobs, such as care for the elderly and environmental clean-up, must be created.

A simpler lifestyle will teach us to make more responsible decisions both as individuals and as a nation. We must recognize the interconnection between personal choice and changes in our political and economic structures. Many of our current individual choices have been shaped by a multinational corporate structure and commercial pressure. Our permissiveness has made us prey to these influences, and collaborators in the rapid depletion of world resources and economic domination of other peoples.

Simple living is revolutionary; it requires a commitment to change our political and social consciousness. Some who have attained a higher level of social consciousness show impatience with the belief that average citizens can change. However, preserving the democratic nature of our American Revolution requires a gradual and systematic educational program embracing changes in individual and, most importantly, communal lifestyles.

This book suggests 99 steps average citizens can take toward simplifying their lifestyle. They are not meant to overwhelm the reader with facts, nor are they meant to upset or discourage the reader with alien suggestions. They are meant to open possibilities that fit personal circumstances and to direct the reader to further information. Hopefully, these hints will serve as a springboard from which the reader can explore simple lifestyles and create one appropriate to individual needs.

A major problem in researching simple lifestyle materials was not the scarcity, but the overabundance of good information. The brevity of the following hints is in keeping with the theme to offer simple suggestions. The final number of 99 ways is not absolute. As we begin to live more simple lifestyles, other suggestions will be forthcoming.

We travel together, passengers on a little spaceship, dependent on its vulnerable resources of air and soil; all committed for our safety to its security and peace; preserved from annihilation only by the care, the work, and I will say, the love we give our fragile craft.

Adlai Stevenson

I HEATING AND COOLING

INTRODUCTION

Non-renewable energy (coal, gas, oil, and uranium) is used in our homes. Due to diminishing supplies, curtailments, and real or fabricated shortages, energy prices are rising; in no area do we experience this more than in our heating and cooling bills. As shown below, 15.7 per cent of our total U. S. energy budget is devoted to heating (space and water) and cooling (air conditioning and refrigeration). When we search for ways to simplify our lifestyles, there is no better place to start.

Conservation-conscious people have two ways of coping with heating and cooling costs: 1) switch to alternative energy sources such as sun, wind, and geothermal (heat from the earth); 2) reduce our consumption of energy.

We can maximize efficiency by insulating homes and offices. Keeping our heating and cooling systems in proper working order will also cut costs. We can operate energy-wasteful appliances only when it is absolutely necessary and at non-peak load hours; and everyone should practice conservation measures such as using hot water sparingly.

ENERGY USE
Per Cent of Total U.S. Energy Use
in Residential Sector

Wood	nil	
Coal	nil	
Natural gas	7.3%	
Electricity	7.5%	
Oil	4.4%	
Total	19.2%	
Space heating	11.0%	15.7% (Space heating through Refrigeration)
Water heating	2.9%	
Air conditioning	0.7%	
Refrigeration	1.1%	
Cooking	1.1%	
Lighting	0.7%	
Other (small appliances)	1.7%	
Total	19.2%	

#1 SAVE HEAT AND INSULATE

Accounting for over half of the household energy budget, heating is the largest energy consumer in the home. With this in mind, let's look at a few ways to save in this area. Revising personal habits and conducting repairs are two ways to ensure that heat is not lost from the house. Using auxiliary heating can also save energy.

Most Americans live and work in uncomfortable overheated buildings. Lowering the thermostat and wearing heavier clothes while indoors will not cause discomfort and will save a significant amount of fuel. For each degree set above 70°, 3 per cent more fuel is used. Cutting back to 65° or 68° F. can save an average household up to $38 in gas, $55 in oil, or $110 in electricity each year.[1]

Living in a cooler home is not harmful to human health; on the contrary, it is better for us. High temperatures can cause the air to be dry, depriving the human respiratory tract of moisture and aggravating bronchial and other respiratory problems. Dryness also causes discomfort to the skin, throat, and nose.

The mind is more alert and the body burns more calories in a cool room. When leaving the indoors to venture into the cold outside, the body is better able to adjust to the winter temperatures, thus reducing chills.

New houses should be built and older ones repaired to prevent heat loss from the building. Repair to conserve heat. Begin with a careful inspection followed by measures such as insulating, weather stripping, caulking, or installing storm windows. Heat is lost through walls, windows, cracks, open doors, and the roof. Keeping windows and doors shut and insulating will reduce heat loss. Weather-stripping seals cracks around windows and doors.

The energy savings from insulation retrofit is exemplified by a study of the New York City metropolitan region. By 1985, if just 10 per cent of the 1970 single family units were retrofitted with additional insulation to reduce energy consumption by 27 per cent, the region would save 1.7 million barrels of oil per year for heating alone, worth $28.5 million (at $.40 a gallon).[2]

When insulating, choose the material with the highest efficiency at an affordable price. Efficiency depends on the material's thickness and on the resistance value "R," which is the measurement of the resistance to heat transfer based on the thickness of the material. A large "R" value indicates a good insulating material.

Thermal insulation reduces the heat transfer from warm to cold with lightweight materials that come in four common forms:

1. Batts or blankets are matted mineral, glass, or cellulose, usually enclosed in paper. One side is coated with asphalt or foil and serves as a vapor barrier. The thickness varies from 1 to 6 inches.
2. Loose fill or granulated is in bags that contain mineral, wool, vermiculite, treated cellulose fiber, or granulated polyurethane. The material is poured from the bag and blown into place. This type is commonly used to insulate ceilings. Granulated forms can be poured into masonry blocks and can reduce heat transfer through walls.
3. Rigid boards are generally used against walls but are less effective than batts or blankets with the same thickness.
4. Reflective insulation is made from foils or polished metallic flake on reinforced paper. This requires about an inch of air space in which the reflective surface hinders heat transfer across the space.

Insulation will pay for itself in a few years by reducing home-fuel consumption and requiring smaller and less costly heating equipment. A consideration to bear in mind is the material's resistance to fire, vermin, and moisture. Wet insulation is virtually useless.

Recommended "R" values for ceilings is 20; walls 12; basement walls 7 to 12; floors without basement 12; floors with unheated basement 7 to 8.

MEASURES TO PREVENT HEAT LOSS[3]

- Locate drafts with a homemade draft gauge; fasten a piece of light cloth over the cross bar of a metal clothes hanger; hold in front of suspected area; if the cloth flutters there is a draft. Drafts can lose 15 to 30 per cent of the heat.

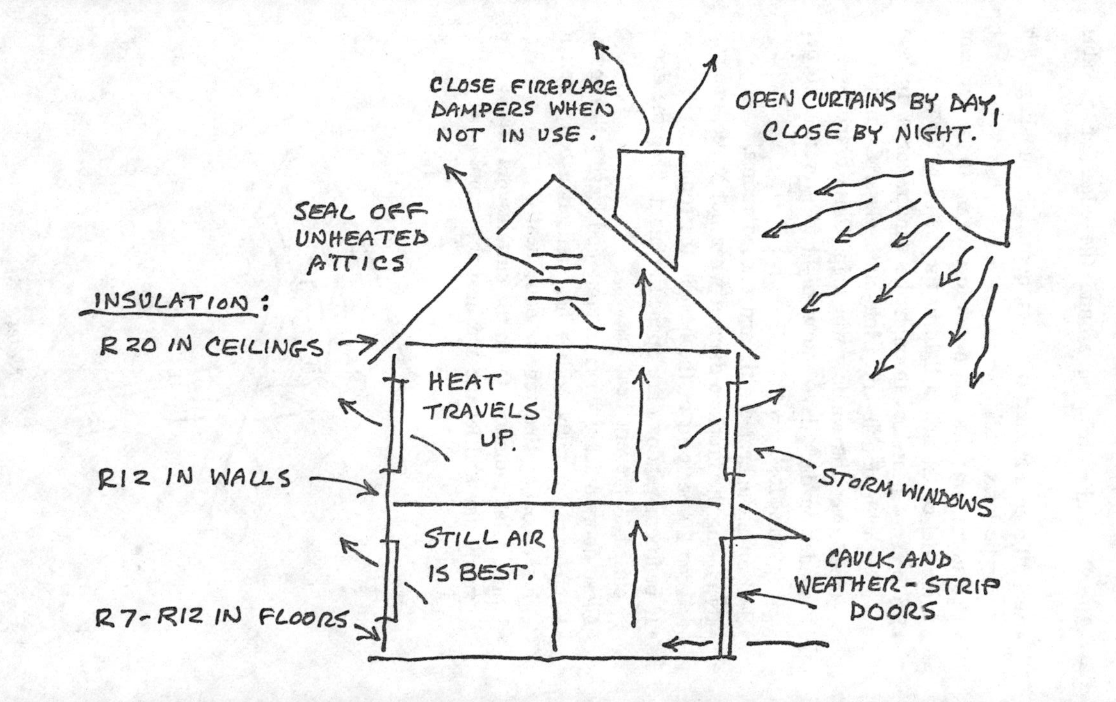
CLOSE FIREPLACE
DAMPERS WHEN
NOT IN USE.
OPEN CURTAINS BY DAY,
CLOSE BY NIGHT.
SEAL OFF
UNHEATED
ATTICS
INSULATION:
R 20 IN CEILINGS
HEAT
TRAVELS
UP.
R12 IN WALLS
STORM WINDOWS
STILL AIR
IS BEST.
CAULK AND
WEATHER-STRIP
DOORS
R7-R12 IN FLOORS

- Install weather-stripping around the edges of movable openings.
- Seal all cracks around windows, doors, and moldings with acrylic latex caulking.
- Use storm windows and doors to save up to 20 per cent on fuel costs; have a tight fit for the best effects.
- To prevent heat loss through the windowpanes, replace glass with double- or triple-plate glass. Clear plastic taped over the window will help. Close draperies and shades on dark days and at night to prevent a 16 per cent heat loss.
- Seal off an unheated attic and unused rooms.
- By closing the fireplace damper when not in use, one can keep 20 per cent of the heat in the room.
- If bedroom windows are opened at night, keep the doors closed to prevent heat loss.
- Close doors quickly when entering and leaving the house.
- Keep warm at night with a heat reflecting bed sheet; it has a coating that reflects body heat.
- Utilize solar radiation. Open the curtains and drapes on a sunny day and let the sun heat the room.

Notes

1. "Energy Saving Tips Given Public," by Paul Hodge, the Washington *Post,* November 11, 1973.

2. "Energy Conservation in Buildings: The New York Metropolitan Region," by Jeffrey C. Cohen and Ronald H. White, Environmental Law Institute, 1346 Connecticut Ave., NW, Washington, DC 20036, July 1975.
3. *100 Ways to Save Energy and Money in the Home,* Canadian Office of Energy Conservation, Creative Generation Ltd., Toronto, Can., 1975.

Additional Sources

"Insulating to Save Energy in Heating, Cooling a Home," by Cecil D. Wheary, *1974 Yearbook of Agriculture Shopper's Guide,* U. S. Department of Agriculture, U. S. Government Printing Office, Washington, DC 20402.

"Save Energy: Save Money!" by Eugene Eccli, *National Center for Community Action,* 1711 Connecticut Ave., NW, Washington, DC 20009.

Low-Cost, Energy-efficient Shelter, by Eugene Eccli (ed.), Rodale Press, Inc., Emmaus, PA 18049, 1975.

#2 KEEP FURNACE OPERATING PROPERLY

A number of home-heating systems are available. Be wise and select the most efficient and economical, based on the size of the house. For safety and effectiveness, the unit must be properly installed, maintained, and operated. A furnace tune-up can reduce home-fuel consumption by 10 per cent or more. For a house that uses 1,500 gallons of heating oil, this is a savings equivalent to $57 each year.[1]

For efficient home heating, keep the furnace, ducts, and register clean and repaired. Clean the inside of the furnace with a vacuum cleaner. Lubricate by filling the oil cups twice a year and keep the fan belt tight. Replace the belt if it is cracked, noisy, or frayed.

For an oil furnace, be sure the firebox and burner nozzle are operating properly. To obtain good burning, install a draft stabilizer and be certain the chimney is clear. Clean the chimney from the top by lowering and raising a brick- or straw-filled sack tied to a rope.

OIL FURNACE CARE

- Remove soot from the firebox, heat exchanger, and pipes.
- Measure the stack temperature. The gases leaving the furnace should be 300° to 450° F. If over this, reduce by increasing the speed of the circulating fan or by reducing the size of the nozzle.
- Observe the smoke density. If there is too much soot, the burner may not be working properly.
- Check the air draw through the firebox and stack; the barometric damper may have to be adjusted.
- Both the furnace and burner should be checked and cleaned twice a year.

When purchasing a gas furnace, get one with an automatic flue gas damper; it will reduce the heat loss when the furnace is not operating. Gas furnaces should be checked annually by a serviceperson to clean the main burners, adjust the flame, clean and adjust the pilot burner, and to check the pilot safety cut-off. To relight the pilot, set the thermostat at a minimum setting, turn on the pilot, and relight.

If buying an electric furnace, get a heat pump type. It uses outside air for heating and cooling and can cut electrical use by 60 per cent.

Heat is lost when it passes through ducts. Therefore, the shortest duct system is the most efficient. Maintain heat by having insulated ducts with caulked joints. Allow air to circulate around radiators, convectors, and registers. Keep them clean and uncovered.

To obtain balanced heating throughout the house, open all vents and read the temperature of each room. Partially close the registers in the hottest rooms. Read temperatures again and readjust the registers to desired heating.

Set the thermostat at 65° to 68° F. and reduce this reading by 6 to 10 degrees at night. Check and clean the thermostat regularly. It should be located away from heat sources, lamps, radios, and television sets, so that it gives a true reading of the house temperature.

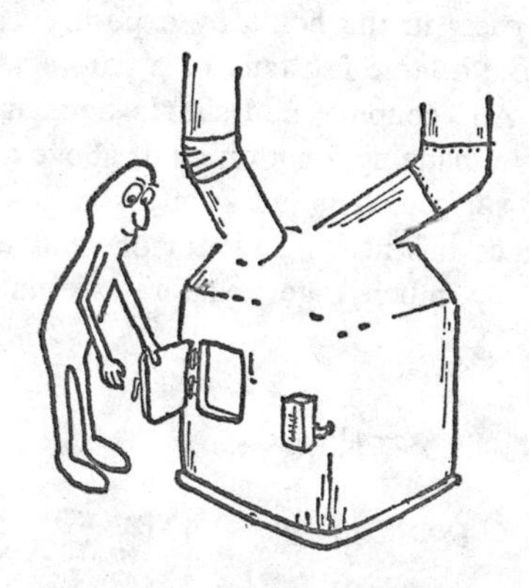

Notes

1. *Energy Reporter,* Federal Energy Administration Citizen Newsletter, October 1975, p. 3.

Additional Sources

The Complete Handyman's Do It Yourself Encyclopedia, by the editors of Science and Mechanics, Stuttman Company, Inc., New York, NY 10016.

"Home Heating Systems/Fuels/Controls," a booklet by the U. S. Department of Agriculture, Superintendent of Documents, U.S.G.P.O., Washington, DC 20402.

100 Ways to Save Energy and Money in the Home, Canadian Office of Energy Conservation, Creative Generation Ltd., Toronto, Can., 1975.

#3 REGULATE HUMIDITY IN HEATED HOMES

The air in centrally heated homes and buildings is often low in humidity, and dryness can be felt on the skin, nose, and throat. Furniture is dry and apt to crack and there is static electricity on rugs and curtains. An average room needs at least a gallon of water to moisten the air for greater comfort. This can be partially attained by using simple methods and totally attained with a humidifier.

Prevent dryness in the home by exposing water to the air for evaporation; a large fish tank or a pan of water will serve this purpose. An economic and simple homemade humidifier can be made by hanging a cotton cloth above a pan of water, touching the surface. The wick effect will keep the cloth moistened. Place this near a radiator or ventilating system. It is surprising how much water evaporates each day.

HOMEMADE HUMIDIFIER

Another way to increase humidity is to leave the water in the tub after bathing and open the door to let the air dissipate. If a dishwasher is used, crack the door after the final rinse cycle. Besides increasing the humidity, energy is saved by letting the dishes air dry.

If health reasons require a more efficient humidifier, an automatic unit can be attached to a hot-air furnace for $200 to $300. Portable humidifiers cost roughly half as much. The best units have a humidistat that cuts off automatically at the desired moisture level.

Most humidifiers operate with a porous fiber or foam filter that obtains moisture from a water reservoir. A fan forces dry air through the wet pad, moisturizes it, and spreads it throughout the house.

OPERATING THE HUMIDIFIER[1]

- Read the owner's manual.
- Place the unit near a room partition and allow 6 inches behind it for circulation.
- Avoid small rooms, heat sources, and corners. The bottom of a stairway is a good location.
- For the first few days, adjust at high settings to moisten air, walls, and furniture. Turn down as dampness develops.
- Set humidistat at a level for personal comfort.
- Disconnect the unit when filling or cleaning the water reservoir. Be sure the area and unit are dry before reconnecting the cord.
- Do not pour hot water in the reservoir or overfill it.

MAINTENANCE[2]

- Dust the grilles every week with a soft brush.
- Every few weeks, empty the reservoir and scrub the inside with a sponge and a mild detergent to prevent mold, mildew, and bacteria.
- Clean the filter pad monthly.

Notes

1. "Portable Humidifiers," G.S.A. Consumer Information Series #17, Superintendent of Documents, U.S.G.P.O., Washington, DC 20402.
2. Ibid.

Additional Sources

House Plants for the Purple Thumb, by Maggie Baylis, Scribner Books, New York, NY, 1973.

#4 CONSERVE COOKING ENERGY

The cooking stove accounts for about 6 per cent of the average household energy used. To reduce the energy used in the kitchen, a number of steps can be taken, ranging from alternative ways of cooking and preparing food to simple conservation measures.

It isn't necessary to eat three hot meals a day. Such foods as meats, salads, cheeses, vegetables, fruits, and soups can be just as appetizing and are often more nutritious when served cold. Meal selections with this in mind are often much easier to prepare, especially for people who return home tired and hungry in the evening after a hard day's work. It will save time and energy to prepare and cook two meals at one time, storing half in the refrigerator for another day.

In preparing hot meals, save electricity and gas while having more fun by cooking out of doors on a charcoal grill or by using a fireplace. Charcoal cooking adds flavor to foods that cannot be imitated by using appliances. Corn and potatoes can be buried in the coals while barbecuing meat or slow cooking a pot of stew. When cold, wet weather forces everyone indoors for warmth and comfort, use the fireplace as a stove. There is no better time for family togetherness.

When using the kitchen stove to prepare meals, a few steps can be followed to conserve energy or to utilize it more efficiently. Have a gas technician turn off the pilot light, which is nothing more than a "dripping faucet" of gas. Instead of constantly burning pilot lights, new gas ranges have electric igniters that can save as much as 30 per cent of the gas normally required.[1]

Since the oven is the heaviest energy user of the stove, use it sparingly. When baking, try to use the oven to cook the entire meal—breads, stews, meats, casseroles, and vegetables could be cooked simultaneously.

When purchasing an electric stove, avoid models with self-cleaning ovens which require more energy. In terms of energy savings, a microwave oven cooks food more quickly and uses 20 per cent less electricity than a regular oven. The cooking heat is generated by vibrating molecules within the food. Microwave oven safety has been questioned, however, because high energy leakage could occur when the closing mechanism on the door is faulty.

Do not preheat the oven for more than ten minutes and only when it is essential. Keep the oven door closed—at least 25 per cent of the heat escapes each time the door is cracked. To fully utilize the retained heat, turn off the oven before the cooking time ends.

Do not block the oven's ventilation with aluminum foil. If placed directly under the pan, foil reflects heat away from the pan and the cooking time is increased. Use a toaster or small grill instead of the oven broiler, since the oven uses 300 per cent more energy.

Thaw meats and meat dishes before cooking—a defrosted roast uses 30 per cent less energy than a frozen one. Foods in covered dishes generally cook more quickly and retain more flavor. Baking potatoes sliced lengthwise cuts cooking time.

Keep the stove burners clean and turn them on only when cooking. Match the size of pots and pans to the burner to avoid wasting heat. Use flat bottom pans that fit the burners and cook with lids on the pots to retain heat. A pressure cooker will reduce cooking time by two-thirds, especially when preparing beans, lentils, and rice. The shorter cooking time also preserves more of the vitamins and minerals in the food.

PRESSURE COOKER HINTS

- Before using a pressure cooker, read the instructions well.

- Keep the lid gasket and pot rim oiled to prevent sticking and create a better seal.
- Do not place food above the recommended level. This will prevent the pressure release valve from clogging.
- Lower the heat when the gauge begins to rock. (Begin measuring the cooking time.)
- Vegetables cook in less than 10 minutes, soaked beans in 45 minutes, stews in 30 minutes.

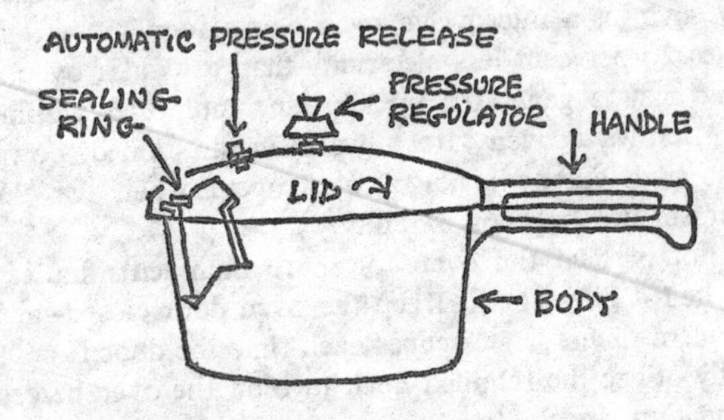

PRESSURE COOKER

No judgment is made here of the relative efficiencies of gas and electric stoves. Although a gas appliance uses more energy than an electrical one in a similar operation, the added energy used to generate electricity makes gas more efficient. The average American electric generation and transmission system must use 3.67 Btu's of energy for every 1 Btu delivered to the appliance. Gas uses only 1.4 Btu's of energy for every 1 Btu delivered.

Price controls and local availability will determine the actual operating costs; however, from an energy conservation viewpoint, gas should be used instead of electricity. With this in mind, the energy savings would be 29 per cent for the cooking range. For other uses gas savings would be: heating–55 per cent; water heater–50 per cent; clothes dryer–58 per cent.[2]

Notes

1. *Energy Reporter,* FEA Citizen Newsletter, Washington, DC 20461, October 1975.
2. "Electricity Demand: Project Independence and the Clean Air Act," by the National Science Foundation, Oak Ridge National Laboratory, Oak Ridge, TN, November 1975.

Other References

"Saving Energy in the Kitchen," by Marion Burros, the Washington *Post,* October 17, 1974, p. F–16.

"A Consumer's Guide to Efficient Energy Use in the Home," a Consumer Affairs Leaflet by the American Petroleum Institute, 1801 K St., NW, Washington, DC.

"Energy Labeling of Household Appliances," National Bureau of Standards, Room A-600, Washington, DC 20234.

#5 REFRIGERATE WISELY

Refrigeration rates with cooking in residential energy consumption, accounting for 6 per cent of the energy used in the home. In the past, ice boxes, cellars, and outdoor storage houses kept food cool and consumed no energy. These methods could be used today for all or part of home refrigeration needs. Beverages can be put on the porch during cold weather, provided they are taken indoors when temperatures drop below freezing. Since the refrigerator is now an accepted part of American life, one should concentrate on ways to best utilize the refrigeration unit, while consuming as little energy as possible.

BUYING

- Don't consider frostless models; they cost up to 30 per cent more, use up to 50 per cent more energy to operate, and are more expensive to maintain than standard models.
- Before buying any model, check the insulation quality. Poor insulation causes the unit to constantly run, wasting energy and reducing the life of the motor.
- The food freezer becomes a good financial investment only if most of a family's food is produced at home. Freezing and storing food can add $.10 to $.25 per pound to food costs. Energywise, a freezer is extremely wasteful, especially uprights (chest freezers use less energy). A well-stocked freezer will require less energy to run than a partially filled one.

LOCATION

- The refrigerator or freezer's location will affect its performance. Leave enough space around the unit for good circulation to remove excess heat. Keep both appliances away from the stove, heating vents, and direct sunlight.

OPERATION

- Empty the refrigerator and turn it off when the house is unoccupied.
- A setting of 40° is more than adequate to preserve food. Check the setting with a thermometer and adjust it to no lower than 40° F.
- Keep dust off the coils in the back of the units; it will increase the efficiency and life-span because dirt acts as an insulator and makes the compressor work longer to maintain the proper temperature.
- Frost accumulation increases energy consumption; defrost when it reaches ¼″ thick.
- Be sure that gaskets around the doors are intact and form

a tight seal. To check, close the door on a piece of paper. If the paper slides out easily, new gaskets may be required.

- Cool down hot foods before refrigerating.
- Don't crowd the shelves. Leave room for the cool air to circulate.

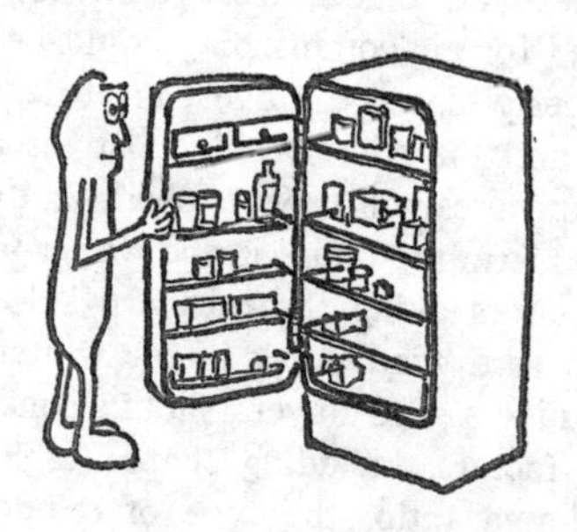

References

1974 Yearbook of Agriculture, U. S. Department of Agriculture, Superintendent of Documents, U.S.G.P.O., Washington, DC 20402.

The Energy Miser's Manual, by William H. Manell, The Grist Mill, Eliot, ME 03903, 1974.

"Now! Energy-Saving Appliances," *Mechanix Illustrated,* Vol. 71, January 1975, pp. 64–65.

Energy Management Digest: "Kitchen Tips," *Commerce Today,* February 17, 1975, p. 12.

AHAM Directory of Certified Refrigerator–Freezers, from Association of Home Appliance Manufacturers, 20 North Wacker Dr., Chicago, IL 60606.

#6 DESIGN BUILDINGS WITH ECOLOGY IN MIND

A substantial decrease in energy use can result from improved building construction and better heating, ventilation, and cooling design. Maintaining comfortable buildings demands about one-fifth of all energy used in the United States. Improvements in existing structures could cut energy consumption without sacrificing service or comfort. If new buildings were erected with energy conservation designs, the energy expense could be cut in half.

Builders should try to situate the structure to minimize the sun's heating effect in the summer and maximize it in the winter. If homes and offices were built to utilize cooling breezes, the need for air conditioning could be cut by 30 per cent. Trees properly placed can assist in protecting a dwelling from the wind and the sun. Deciduous trees on the eastern and southern exposures will provide shade in the summer and allow the sun to warm the house in the winter when the leaves have fallen. Conifers along the northern side give shade in summer and act as a wind break in the winter.

Architects, builders, and buyers should consider a number of construction factors, including shape (high rise or low), windows-to-wall-area ratio, the type of windows and glass, building materials, and insulation.

Tall buildings suffer a greater heat loss than low ones. Hot air rises to the top floors and is lost. Cold air is sucked in the bottom floors by low pressure and must be heated. Therefore, this effect wastes the building's heat by requiring this constant supply of cold air to be heated. A low building corrects this problem by allowing a broader and more efficient dispersal of air.

BUILD TO CONSERVE HEAT IN WINTER AND COOLNESS IN SUMMER[1]

- Cover the exterior walls with earth and plantings.
- Shade walls to reduce indoor/outdoor temperature differential. Shade paved areas and plant vegetation to lessen outdoor temperature buildup.
- Locate building to minimize the wind's effects on exterior surfaces. Use sloping sites to protect the building.
- Select a site to reduce heat reflections from water.
- Construct with minimum exposed surface area to minimize heat transmission. A minimum north wall will reduce heat loss, and a minimum south wall will reduce cooling load.

- Use a building configuration and wall arrangement that provides self-shading and wind breaks.
- Construct exterior walls, roof, and floors with a high thermal mass.
- Provide a vapor barrier on the interior surface of exterior walls and roof of sufficient impermeability.
- In a warm climate finish outside walls and roof with a light-colored surface to minimize heat gain. In a cold climate use a dark-colored finish that has high solar absorption.

Note

1. "A Checklist of Energy Conservation Opportunities, Ranked in Priority According to Climatic Conditions," *A.I.I. Journal*, May 1974, American Institute of Architects, 1735 New York Ave., NW, Washington, DC.

Additional Sources

Shelter and Society, by Paul Oliver, Praeger Publishers, New York, NY, 1969.

Your Engineered House, by Rex Roberts, Lippincott Co., Philadelphia, PA, 1964.

Architecture Without Architects, by Bernard Rudofsky, Doubleday & Co., Garden City, NY, 1964.

"Earth for Homes," U.S.D.A. Publication #PB188918, Superintendent of Documents, U.S.G.P.O., Washington, DC 20402.

The Wilderness Cabin, by Calvin Rutstrum, Macmillan and Co., New York, NY.

Stone Shelters, by Edward Allen, MIT Press, Cambridge, MA 02142.

#7 CUT HOT-WATER COSTS

Heating water is the second greatest energy consumer after home heating. The average home water heating use ranges from 23 to 41 million Btu's per year. To save energy in heating water, consider the hot-water tank in terms of its size, type, and location. Repair faulty plumbing and correct bad habits that cause hot water to be wasted.

The tank should have a thick insulation shell. Its recovery rate should be at least 75 per cent, which means that a 40 gallon tank will reheat 30 gallons in one hour, which is sufficient for average family use. If the control is set at 120° F., it will give sufficient hot water. The heater should be located in a warm place close to the kitchen so that the hot water pipes are short and the heat loss is diminished. The system can be insulated by wrapping fireproofed insulation strips around the pipes. The insulation can be held in place with tape.

Keep the tank in condition by servicing it once or twice a year. Check the flame, barometric damper, stack temperature, and make sure the energy is used to heat the water and is not being dissipated up the chimney. Once a month, open the tap at the bottom and drain enough water to remove sediment and mineral deposits.

Bath water can be used for heating purposes in winter by leaving hot bath water in the tub until water reaches room temperature.

A number of steps can be taken to reduce hot water bills.

- Prevent drips. One drop per second accumulates to 175 gallons per month which represent a significant addition to the annual fuel bill.
- Take showers of short duration and use less water while tub bathing. Showers generally require less water than tub baths. When the shower is too warm, turn down the hot water instead of increasing the cold.
- Wash dishes by hand. Start with a small amount of water in the sink and gradually add more water as you wash

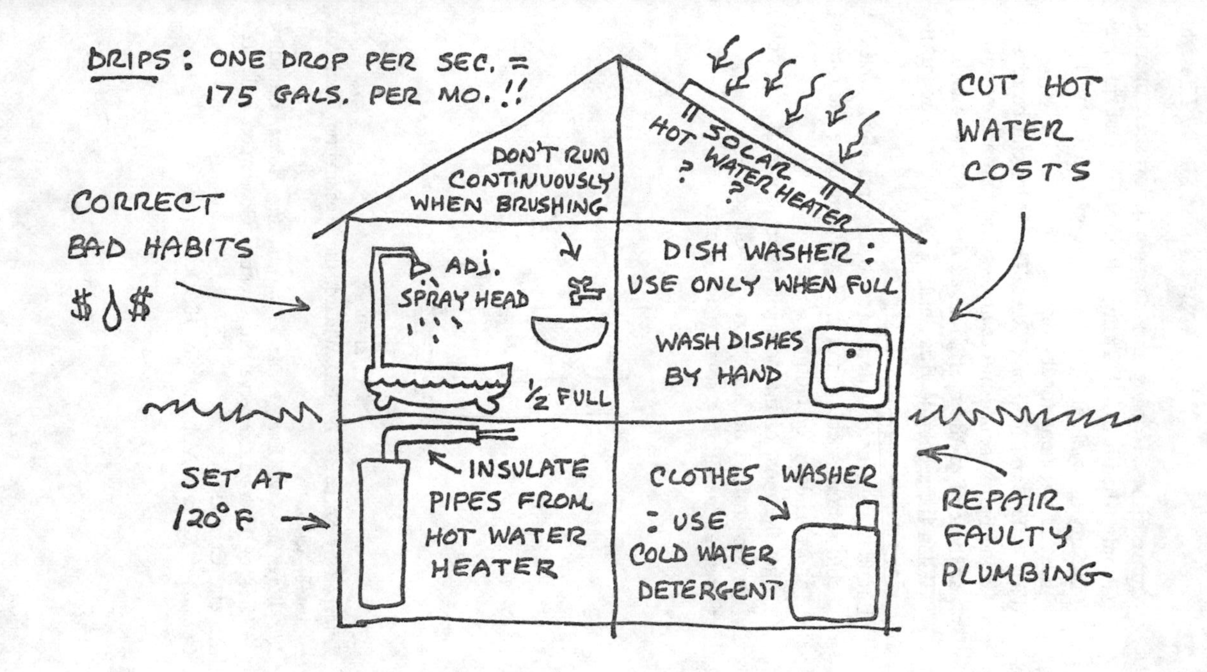
DRIPS: ONE DROP PER SEC. = 175 GALS. PER MO.!!
CUT HOT WATER COSTS
DON'T RUN CONTINUOUSLY WHEN BRUSHING
SOLAR HOT WATER HEATER ? ?
CORRECT BAD HABITS
$ $
ADJ. SPRAY HEAD
DISH WASHER: USE ONLY WHEN FULL
WASH DISHES BY HAND
½ FULL
INSULATE PIPES FROM HOT WATER HEATER
SET AT 120°F
CLOTHES WASHER: USE COLD WATER DETERGENT
REPAIR FAULTY PLUMBING

more dishes. If using an automatic dishwasher, operate it only when it is full. Turn it off after the final rinse, as the drying cycle is a waste of energy and the retained heat is sufficient to dry dishes.

- Wash only a full load of clothes. The water can be reused if the lightly soiled pieces are washed first and removed at the end of the cycle. Reload and reset the machine. When the second load is finished, the first can be put in for rinsing.
- Launder with cold water. Most detergents clean well in cold water. A cold-water wash will minimize the fading and shrinking of clothes.
- To get the most of the dollar invested in energy, use cold water instead of hot whenever you can.
- Consider installing a solar water heater.

References

"Flushed by Success," by Hal Willard, the Washington *Post,* August 8, 1975.

100 Ways to Save Energy and Money in the Home, Canadian Office of Energy Conservation, Creative Generation Ltd., Toronto, Can. 1975.

Consumer Reports, Vol. 39, No. 10, October 1974.

#8 ECONOMIZE WITH MODULAR HEATING

In addition to doing everything possible to prevent heat loss from the house, it is essential to have the most economical heating for the particular type of dwelling and climate. For example, for a small house or apartment in an area with mild winters, a heating unit such as a wood stove, coal stove, fireplace, gas heater, oil heater, propane heater, or electric space heater can be more economical than a central furnace.

When choosing a space heater, keep safety and energy use in mind. Electric heaters should be considered last because they are heavy users of energy. Non-electric heaters should be properly vented. If venting is inadequate or the adjustments faulty, deadly quantities of carbon monoxide may be produced. Many states prohibit unvented units because of this hazard.

A space heater should have a guard around the flames or coils to protect children, pets, and clothing from the heat source. Allow adequate clearance on all sides of the unit.

To keep the heater in good working order, inspect it regularly for needed adjustments, cleanliness, cracks, faulty legs, and hinges. Keep the electrical wiring in good condition.

FOR SAFE OPERATION OF SPACE HEATERS[1]

- Keep a window partially open if an unvented unit must be used. Fresh air will prevent the accumulation of gas fumes.
- Use only the fuel the heater is designed for. Do not convert to another fuel without consulting an expert.
- Keep children away from space heaters and stoves.
- Be certain the fire has proper ventilation to maintain a constant rate of burning that gives moderate heat, not too hot or too cool.
- Always keep a screen around a heater that has open flames.

- Keep the damper open while the fuel is burning. This will provide for efficient burning and will prevent accumulation of explosive gases.
- Learn how to light a gas heater properly. If someone smells gas, turn off all controls and open a window. Don't allow gas to accumulate. If heater fails to light on the first try, allow sufficient time for the gas to dissipate before trying again.
- Never keep flammable liquids around heaters. Vapors can be ignited by the open flames.
- Use heavy-duty cord for electric heaters. Have an electrician check the wiring if a heater uses higher than usual wattage.
- Never place an electric heater near a bathtub, shower, or sink. Don't touch one when wet.
- Place a metal sheet under the unit to protect the floor from live coals, burning, or overheating.

The fireplace is used today primarily as an enjoyable addition to a house. Watching the flames can bring peace to the heart and soothe the soul. Even more importantly, however, the fireplace can be used to supplement central heating. On cool days, the fireplace might provide the necessary heat to take away the chill.

The most energy-efficient fireplaces will have warm air vents through which hot air can be spread around the room or through ducts to other rooms. A modified fireplace with a fan can also help circulate hot air effectively.

To obtain maximum efficiency from a fireplace, be sure the chimney is working as it should. If the chimney is too low, the draft will not be strong enough to pull out the smoke. Keep the chimney clear of ashes and be sure the flue is operating properly.

An improperly built fire can lose 90 per cent of its heat up the chimney. Physicist Lawrence Cranberg has designed a fire grate that forces the fire to burn and distribute heat more efficiently. An additional bonus is that the fire is easy to light. (See diagram.) For information write: 1205 Constant Springs Dr., Austin, TX 78746.

A prefabricated fireplace can be easily installed. They are less expensive than masonry fireplaces but are usually less durable.

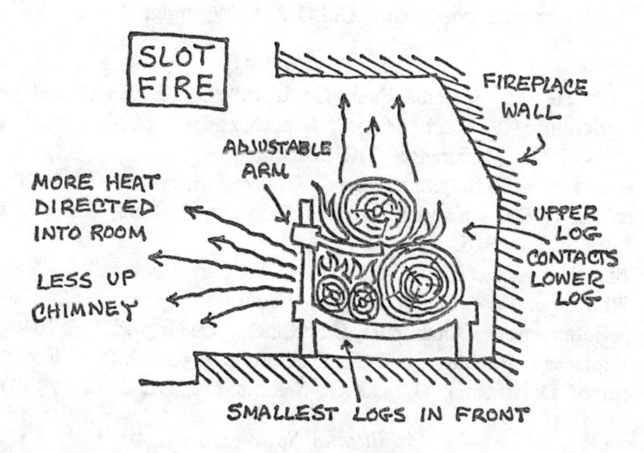

Wood, coal, and natural gas can be burned in fireplaces with these fuels varying in their heat production and efficiency. As an alternative heating supply, tightly rolled newspapers can be burned as "logs." Tie a string or wire around the end to keep it fastened. Soaking the log in water with a spoon of detergent added to it will reduce the amount of ashes when the logs are burned. Dry the logs thoroughly before using them.

When the fire is out, be sure to close the damper, or the heat will escape up the chimney. If the fireplace is never used, prevent heat loss by sealing it off and plugging the chimney.

Note

1. "Space Heaters and Wood and Coal Burning Heating Stoves," Consumer Product Safety Commission Fact Sheet #34, Superintendent of Documents, U.S.G.P.O., Washington, DC 20402.

Other References

"Home Heating Systems/Fuels/Controls," U. S. Department of Agriculture Bulletin #2235, Superintendent of Documents, U.S.G.P.O., Washington, DC 20402.

The Complete Handyman's Do It Yourself Encyclopedia, by the editors of Science and Mechanics, Stuttman Company, Inc., New York, NY 10016.

"Cozy Fireplaces, Franklin Stoves and Other Heaters," by Vera Ellithorpe, *1974 Yearbook of Agriculture Shopper's Guide,* Superintendent of Documents, U.S.G.P.O., Washington, DC 20402.

"Fireplaces and Chimneys," U.S.D.A. Bulletin #1889, Superintendent of Documents, U.S.G.P.O., Washington, DC 20402 ($.40).

Additional Sources

Observations on the Forgotten Art of Building a Good Fireplace, by Vrest Orton, Yankee, Dublin, NH.

#9 COOL CONSERVATIVELY AND VENTILATE

It is possible to be comfortable during summer without the expense and energy waste that goes with air conditioning. High heat and humidity are conditions that can be dealt with by taking a few prudent steps to improve personal comfort and state of mind.

Keep cool by wearing loose-fitting, lightweight, and light-colored clothes. The grammar school science teacher was correct–light colors reflect heat and dark shades absorb it. Cold drinks are cool and refreshing, so relax and take a break during the stifling parts of the day. A cool shower may help also, or just soak in a tub or pool.

Reduce heat generation by not operating unnecessary appliances. Perform housework–vacuum, iron, wash clothes–in the cooler parts of the day, either early morning or late evening. Avoid using the stove or oven to prepare meals–eat salads, chilled fruits, cold cuts, and cheeses.

Turn off unnecessary lights, especially incandescent bulbs. Fluorescent lights burn cooler, give more light, and cost one-fourth as much to operate. Also, after washing dishes or bathing, don't leave hot water standing; extra heat and humidity is the last thing needed.

Keep the sunlight outside by closing drapes during the day. Light-colored drapes and blinds reflect the sun and can reduce solar heat by 50 per cent. Hang awnings and window shades on the south side and shade the east and west with trees. Deciduous trees will provide abundant shade and block the sun's heat (see entry #41).

To achieve the greatest degree of coolness, it is important to keep the air circulating. Fans–room, window, or attic models–serve this purpose well and may eliminate the costly need for air conditioning.

Room fans are available in various sizes from table to floor models. The most effective ones have large blades that revolve up to 1,000 revolutions per minute. Energy use, speed adjustments, oscillating mechanisms, and noise levels are other points to consider when purchasing a room fan.

At night, utilize cool outside temperatures by opening the windows. Attic and window fans can be used to pull this cool air in and push warm air out. Attic temperatures can reach levels 25 degrees warmer than the outside air and thus can create a broiler effect in the house. By exhausting trapped heat attic fans serve to cool the entire home. A good attic fan is equipped with louvers that open when the fan is on and remain closed at other times.

Window models, much easier to install than attic fans, also draw on outside air to increase inside circulation and reduce heat levels.

Once again, insulation can make a big difference in home-cooling comfort. Insulation helps keep heat out in summer, following the adage that an ounce of prevention is worth a pound of cure. Refer to entry #1 for information on insulating.

References

100 Ways to Save Energy and Money in the Home, Energy, Mines and Resources, Canadian Office of Energy Conservation, Creative Generation Ltd., Toronto, Can. 1975.

Public Works—A Handbook for Self-Reliant Living, edited and compiled by Walter Szykitka, Links Books, New York, NY, 1974.

#10 DEHUMIDIFY IN SUMMER

Summer discomfort is largely caused by high humidity. The air's excess water content prevents sweat from evaporating and thus diminishes the body's natural cooling effect.

Removing excess water from the air in homes is possible without resorting to the expense of air conditioning. A dehumidifier is an apparatus that draws room air over a cooling coil where it loses moisture. Water is removed by condensation on cold coils. It is then heated slightly and returned by fan.

When buying a unit, look for the water removal capacity rating of the Association of Home Appliances Manufacturers.

Dehumidifiers can range in price from $90 (with a water removal capacity of eleven pints a day) to $150 (with a 30 pint capacity and convenience features). The electrical operating cost can range from $.20 to $.50 per day. Operating costs are reduced when an area has been sufficiently dried and the machine does not have to run steadily.

FOR SAFE, EFFICIENT OPERATION[1,2]

- Read the owner's manual.
- Place the unit at least 6 inches (15cm) from the nearest wall for free air flow.
- Shut all doors and windows in the area to be dried.
- For the first few days, turn the humidistat to "Drier" or "Extra Dry."
- Turn the machine off and disconnect power cord before emptying the water pan; be sure the area and unit are dry before reconnecting the cord.
- On cool days, check the cold coils for frosting; turn off until melted.
- Always unplug the cord before cleaning the unit; dust the grills and wipe the cabinet.
- Every few weeks, clean the inside of the water container with a soft cloth and a mild detergent to prevent mold, mildew, and bacteria.
- Once a month, dust the lint from the cold coils.

Notes

1. *Handbook for the House,* Consumer Information Series, Public Documents Distribution Center, Pueblo, CO 81009.
2. *Portable Dehumidifiers,* G.S.A. Consumer Information Series #77, Superintendent of Documents, U.S.G.P.O., Washington, DC 20402, 1974.

#11 AIR CONDITION MINIMALLY

In 1950, air conditioning was present in less than 1 per cent of all American households. By 1974, 48.6 per cent had either room or central air conditioners. Along with this surge in consumer purchases have been higher electricity costs accompanied by frequent brownouts and peak load power problems. Residential energy consumption for air conditioning jumped six-fold from 1960 to 1972.[1]

Air conditioning is an expensive luxury that depletes electrical power resources. Therefore, users should stress conservation and efficiency measures when operating their units and use them as few times as possible. Entries #9 and #10 suggest a few alternatives to air conditioning.

For various reasons, many Americans prefer to pay the high cost of air conditioning. However, there are ways to save energy and be economical when operating this equipment. Insulating the home is a primary step—this single practice can save 20 to 30 per cent on operating costs[2] (see entry #1).

There are two main ways to air condition a house. Room units are window-mounted and are used to cool one or several rooms. Central systems are used to cool an entire house or building. Room units are often preferable because home cooling can be limited to frequently occupied rooms. Keeping doors and windows shut will make selective cooling very efficient. Furthermore, the units can be turned on only as needed, and most models are equipped with various fan speeds and temperature settings that allow for the most efficient cooling.

Central systems consume more energy because they cool a

wider area, much of which is never occupied. There is no need to cool the entire house if only one person is inside during the day. Cool air is lost as it spreads from the refrigeration unit through the ducts. A single thermostat instead of zone control further reduces a central system's cooling efficiency.

A 1970 comparison of operating costs showed that room units consumed an average of 1,935 kilowatt hours per household, while central units used 3,560 kilowatt hours per household.[3] This difference illustrates the additional costs that accompany a central system. With electricity costs rising yearly, this represents a tremendous waste of the homeowner's dollars and the nation's energy.

Regardless of the unit, a temperature setting of 75° to 78° F. will provide adequate comfort. Even at this temperature, units should be operated only when absolutely necessary; instead, open the windows or cool the house with fans.

Simple maintenance procedures will ensure a more efficient unit. Don't let shrubs or leaves block the air flow into the unit. Keep dust and dirt out by cleaning and replacing the filters periodically. The evaporator and condenser also need to be cleaned occasionally.

Operating output is in British thermal units (Btu's) per hour, ranging from 4,000 to 24,000 Btu's. The efficiency of individual units is rated in terms of cooling obtained to the amount of electricity used. Determine the efficiency by dividing the Btu rating per hour by the wattage required. The higher the number, the greater the efficiency and the less electricity used.

The importance of efficient units is illustrated by the fact that the average room air conditioner in 1970 had an efficiency of 6 Btu's per watt hour. If this average was increased by 1980 to 10 Btu's per watt hour, the electricity consumed could be reduced by 40 per cent.[4] Remember, different brands of air conditioners may vary threefold in their efficiency.

Since energy is wasted by oversized units, do not buy a larger unit than is needed. Know the volume of the area to be cooled. Average Btu requirements for rooms vary:

Bedroom–5,000–6,000

Living room–8,000–12,000

Several connected rooms–15,000–20,000
Medium-sized house–24,000

OPERATING SUGGESTIONS[5]

- Set thermostat at 75° to 78° F.
- Keep heat out of the house by closing all windows and doors; seal air leaks with weather-stripping or caulking.
- Keep storm windows in place; pull shades or drapes across windows during the day.
- Cool the attic with a ventilation fan.
- Turn off appliances and lights when not in use.
- Cook main meals, take showers, wash, and iron in the morning and evening.
- Avoid direct sunlight on the cooling system.
- When humidity is high, set the fan at a low speed; this allows more water to be removed.
- Don't cool more rooms than is absolutely necessary. Cool by opening windows if the day is not very hot.

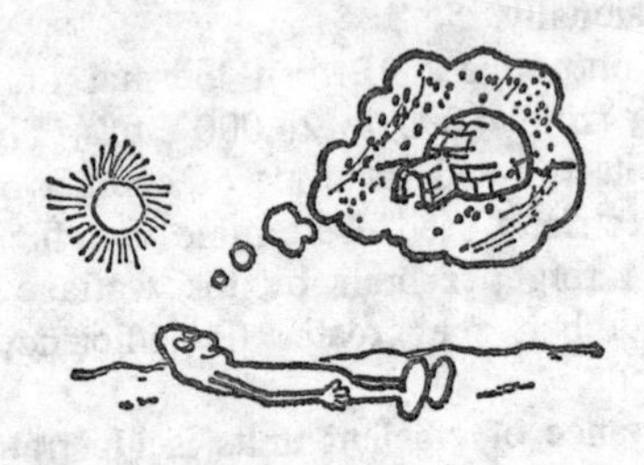

CARE AND UPKEEP[6]

- Unplug before performing home maintenance.
- Clean or change filters at least once a month. Wash in soap and water and replace while damp. If a new one is needed, get the same material, quality, and thickness as the original.
- To prevent rusting, cover in the winter with a moisture-proof cover.
- Vacuum the cooling coils so they are free of dust and lint.

Notes

1. *Potential for Energy Conservation in the United States: 1974–1978 (Residential-Commercial)*, National Petroleum Council, Washington, DC, September 10, 1974.
2. *The Room Air Conditioner as an Energy Consumer,* by John C. Moyers, National Science Foundation Environmental Program, Oak Ridge National Laboratory, Oak Ridge, TN, October 1973, p. 12.
3. *Residential Consumption of Electricity 1950–1970,* by John Tansil, National Science Foundation Environmental Program, Oak Ridge National Laboratory, Oak Ridge, TN, July 1973, p. 11.
4. Ibid., p. 12.
5. *Room Air Conditioners,* G.S.A. Consumer Information Series #6, Superintendent of Documents, U.S.G.P.O., Washington, DC 20402, 1972.
6. Ibid.

Other References

Room Air-Conditioner Lifetime Cost Considerations: Annual Operating Hours and Efficiencies, by David A. Pilati, Oak Ridge National Laboratory, Oak Ridge, TN, 1975.

#12 CONVERT TO RENEWABLE FUEL SOURCES . . . ORGANIC

There are numerous organic and environmental energy sources, some of which have been used since the dawn of civilization. A conservation and energy-conscious generation should search through man's history to rediscover abundant renewable energy sources and, wherever possible, curb consumption of non-renewable ones such as oil, gas, and coal. Some renewable organic sources include: wood, plants, agricultural wastes, and biological wastes (burned in dried form or converted into methane gas).

Early colonists were awed by the "endless" acres of American forests. They took advantage of this inexpensive forest fuel and for several centuries wood remained a primary energy source. One hundred years ago, 75 per cent (2,868 trillion Btu's) of America's energy was wood.[1] Today, wood accounts for only ⅓ of 1 per cent of America's energy.

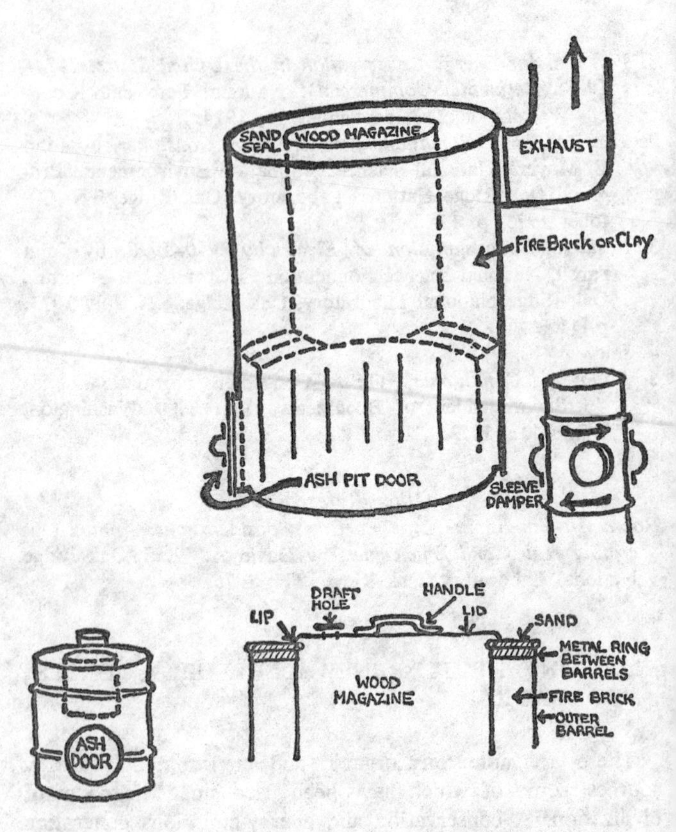

Homemade Wood Stove

Parts for a wood heater: 50-gallon steel barrel; a half barrel; fire brick or other fire-box lining. Cut both ends out of half-barrel. Save the ends. Cut hole in top of full barrel just large enough to accept half-barrel. Drop half-barrel through hole and fasten air-tight to top of full barrel by welding or by using mechanical fasteners and caulking with furnace cement. The half-barrel serves as the wood magazine. Line inside of big barrel with fire brick as shown. Fashion a lid for magazine using an end saved from half-barrel. The lid must fit tightly and have an adjustable draft opening. A flue outlet must be placed at the top back of the big barrel. An ash pit

Wood wastes from lumber and paper mills are allowed to rot or, worse, are buried in landfills. These wastes account for 4.6 million tons (3.7 per cent) of the residential, commercial, and institutional solid waste accumulation.[2] Wood residues—limbs, stumps, partially decayed wood—amount to about 9.6 billion cubic feet a year, almost ¾ as much as is used yearly for wood products.[3] Young trees are often eliminated in thinning operations while others are killed by disease or fires.

Collecting wood wastes is a major social event in European lands where the cultivation and upkeep of forests takes a higher priority than in our country. Part of America's difficulty is that the wastes are generally far removed from the point of consumption, and transporting this wood is a major expense.

If properly managed, a lot measuring 6 to 12 acres can provide enough wood to heat a home indefinitely. (Pound for pound, hardwood has one-half the heat value of coal and one-third that of oil; soft wood is far less.) Millions of scrub acres in urban areas could be better managed, and wood could be removed for burning. Planting good firewood trees could be emphasized.[4] Eucalyptus trees are especially suited for firewood growth since they grow rapidly and survive in crowded conditions. Firewood prices have inflated both in the United States and abroad. If an average American household bought its own firewood, it would cost at least $300 per year.

Using arable land to produce plants for energy production may seem irresponsible and in conflict with our capacity to produce food. However, integrating food and energy production might prove to be worthwhile. For example, the Chinese produce methane by mixing weeds, wood scraps, human feces, guano, and animal droppings in sealed septic tanks.

---

door at the bottom of full barrel is necessary. It serves as access door for starting fires and for cleaning out ashes. To start the fire, work through ash pit door. Use kindling and when good fire is started, add wood to magazine. Fire burns below the magazine in lined part of big barrel. Exhaust escapes between inner and outer barrels and up flue. Fire can be controlled using sleeve damper in exhaust flue and the draft opening in magazine lid. For more information contact: Ted Ledger, 591 Windmore Ave., Toronto, Ont. M6S. 3L9 Canada.

An easy-to-build 10-cubic meter "generator" could supply a family of five with cooking and lighting fuel. An additional advantage to generating methane is that valuable organic residues can be salvaged and spread on the fields as a fertilizer.

Urban organic wastes cannot be properly utilized on a household-by-household basis. However, community programs could comply with the requirements for a simpler lifestyle. One ton of solid waste yields 10 million Btu's. If incinerated to generate electricity, 10 per cent of 1970's solid waste could have yielded 25 billion kilowatt hours (1.5 per cent of U. S. production).[5]

Notes

1. *The Contrasumers: A Citizen's Guide to Resource Conservation,* by Albert J. Fritsch, Praeger Publishers, New York, NY, 1974, p. 18.
2. *Energy in Solid Waste: A Citizen Guide to Saving,* Citizen's Advisory Committee on Environmental Quality, 1700 Pennsylvania Ave., NW, Washington, DC 20036, p. 13.
3. "Feature," from the National Wood Research Center of the Forest Service, U. S. Department of Agriculture, Washington, DC, October 20, 1975.
4. *The Complete Book of Heating With Wood,* by Larry Gay, Garden Way Publishing, Charlotte, VT 05445.
5. *Energy Implications of Several Environmental Quality Strategies,* by Eric Hirst, National Technical Information Service, 5285 Port Royal Rd., Springfield, VA.

Additional Sources

The Mother Earth News Handbook of Home-Made Power—How to use wind, water, wood, methane, and solar energy to power your home, by the staff of Mother Earth News, Bantam Books, New York, NY, 1974.

Bio-Gas Plant: Generating Methane from Organic Wastes, by Ram Bux Singh, Ajitmal, India, Gobar Gas Research Station, 1974. (For a copy write to Mothers Bookshelf: PO Box 70, Hendersonville, NC 28739.)

The Woodburners Handbook, by David Havens, Media House, Portland, ME, 1973.

Methane Digesters for Fuel Gas and Fertilizer, from the New Alchemy Institute—West, Box 376, Pescadero, CA 94060.

Alternative Sources of Energy Magazine, Alternative Sources of Energy, Rt. 1, Box 36B, Minong, MN 54859.

The Energy Primer, from the Portola Institute, 558 Santa Cruz Ave., Menlo Park, CA 94536.

Energy for Survival, by Wilson Clark, Anchor Books/Doubleday, Garden City, NY, 1974.

Alternative Energy Resource Organization
418 Stapleton Bldg.
Billings, MT 59101

#13 CONVERT TO RENEWABLE FUEL SOURCES . . . SOLAR

Besides wood and organic materials, there are other renewable energy sources that should be more fully utilized in an energy-conscious age. Since these are not major polluters —as are high sulfur coal and nuclear fuels (radioactive emissions from power plants)—they are called "environmental" energy sources: solar, wind, geothermal, tidal, and hydroelectric. No major technological impediments bar the use of these renewable sources, and if an effort is made to develop them, they could supply a large portion of our energy needs by the 1990s. They could replace the rapidly dwindling nonrenewable natural gas supply.

Solar energy, which may be the energy wave of the future, can be felt by standing in front of a picture window on a sunny day. Solar heating, and to a lesser degree cooling, is ripe for use on a small scale, but generating electricity from solar sources lies in the future. A few technical hurdles and legal barriers to air rights prevents widespread solar heating and cooling, but these problems are not insoluble. Thousands of solar water and space heaters are currently used in Japan, Israel, and Australia. Solar energy could halve commercial and residential space and water heating costs by 1990 and would save 10 per cent of our country's total energy budget.[1]

DO-IT-YOURSELF SOLAR HEATING TECHNIQUES

- LET THE SUN SHINE IN. When sealed against drafts, windows provide heat in the winter; open the curtains when the sun is shining.
- GRAVITY WARM AIR WINDOW HEATERS. A fiber-glass-insulated plywood box is covered with a sheet of glass over black painted metal. The metal makes two continuous chambers inside the house. The box extends at an angle from below the window toward the ground. Air heated by the sun in the upper chamber creates a circulation that pulls cooler air from the room's floor level. The box can also be used on a roof. (See diagram.)[2]
- REFLECTIVE PANELS TO INCREASE FLOW OF SUN'S HEAT INTO A ROOM. Panels consisting of ⅜-inch exterior plywood, a layer of insulation, and a reflective surface (mylar or aluminum) are held to window frames with adjustable cords. The panels are adjusted to catch sunlight. These panels can be located on upper or lower windows, or one panel could cover both windows.[3]
- FLAT-PLATE COLLECTORS. A box made of glass plates, pipes, insulation, and absorbing surface. On a black surface, glass acts like a greenhouse. The glass transmits short solar light waves to the black surface which absorbs them; the glass keeps the long heat waves from passing out, thereby retaining maximum heat.

  Pipes containing water or air run through the box. These pipes can be part of a circulating system to heat rooms or go to a storage container for future use as a hot-water supply.

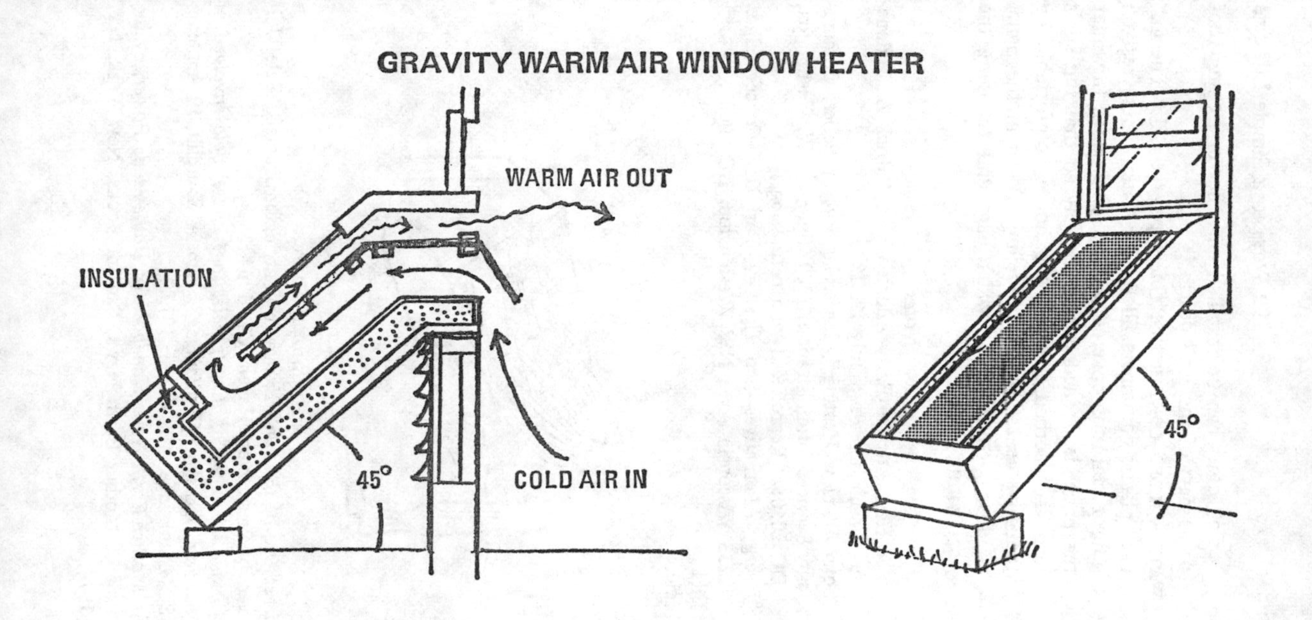
GRAVITY WARM AIR WINDOW HEATER
WARM AIR OUT
INSULATION
45°
COLD AIR IN
45°

Collectors can be installed on a flat or sloping roof or as part of the wall design.[4]

- WATER-FILLED CONTAINERS. One type, the "Sky Therm," uses plastic containers filled with water. These rest on a flat metal roof, covered with a sliding panel. In winter the panel is opened during the day so that the sun's heat is absorbed, keeping the house comfortable by radiating heat through the ceiling. At night the panel is closed, covering the water bags and retaining the heat within them.[5]

Notes

1. "Solar Heat Attracts Early Adopters," by Patrick A. Malone, the Washington *Post,* July 19, 1975.
2. "Solar Window Heating," Conserving Energy Series, Community Services Administration: 1200 19th St., NW, Washington, DC 20506. (Reproduced with permission.)
3. "Energy Alternatives–Eco Tips, #7, Part III," Concern, Inc., 2233 Wisconsin Ave., NW, Washington, DC 20007.
4. Ibid.
5. Ibid.

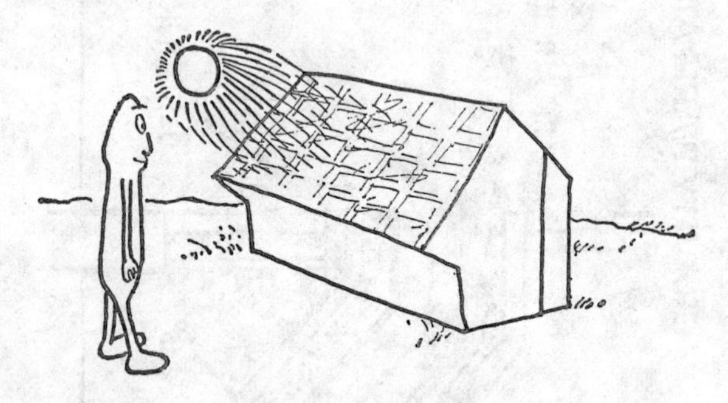

Additional Sources

Direct Use of the Sun's Energy, by Farrington Daniels, Yale University Press, New Haven, CT, 1964.

Informal Directory of the Organizations and People Involved in the Solar Heating of Buildings, by Wm. A. Shurcliff, 19 Appleton St., Cambridge, MA 02138, 1975.

The Mother Earth News Handbook of Home-Made Power, by the staff of Mother Earth News, Bantam Books, New York, NY, 1974.

"List of Solar Manufacturers," *Synergy*, PO Box 4790, Grand Central Station, New York, NY 10017.

"How to Build a Solar Water Heater," from the Environmental Information Center of the Florida Conservation Foundation, Inc., 935 Orange Ave., Suite E. Winter Park, FL 32789.

"Solar Heated Greenhouse Plans," from the Citizens for Energy Conservation and Solar Development, PO Box 49173, Los Angeles, CA 90049.

"Solar Energy Alternatives," by Bettina Conner, Institute for Policy Studies, 1901 Q St., NW, Washington, DC 20009.

Energies Newsletter, from Solar Energy Society, PO Box 4264, Torrance, CA 90510.

Solar Directory, by Carolyn Pesko, Ann Arbor Science Publishers: PO Box 1425, Ann Arbor, MI 48106.

"People and Energy Newsletter," by Ken Bossong, CSPI Publications, 1757 S St., NW, Washington, DC 20009.

Solar Energy-Technology and Applications, by I. Richard Williams, Ann Arbor Science Publishers, PO Box 1425, Ann Arbor, MI 48106.

#14 CONVERT TO RENEWABLE FUEL SOURCES . . . WIND

Wind power is not a novelty. For thousands of years man has used the wind to sail ships. The Dutch pumped water from their soggy marshlands with large windmills that still dot the countryside. English colonists constructed a windmill in Williamsburg, Virginia. The heyday of the American windmill was in the late nineteenth century when it pumped water for livestock troughs and irrigation ditches on Western farms. However, technological "progress," particularly in the form of rural electrification, has shut down many of these windmills.

With a renewed interest in environmentally sound energy systems windmills are going to make a comeback. As many as 50,000 could be repaired and put back in use. Many new and better designed windmills could be constructed especially in the Great Plains, mountain and coastal regions where prevailing winds could supply up to 20 per cent of the total U. S. energy needs (see Environmental Energy Map).

Professor William Heronemus of the University of Massachusetts Department of Civil Engineering has helped revive the large-scale wind-power concept in America. He has proposed an offshore wind-power system that would be capable of generating electricity economically for all of New England to meet current demands. The wind power would be used to convert sea water by electrolysis into combustible hydrogen gas.[1]

Under an Energy Research and Development Administration program, the National Aeronautics and Space Administration is constructing a large wind-powered generator aiming toward developing cost-competitive wind-energy conversion systems. These systems are expected to be in commercial use by the 1980s.

However, great potential still rests with the small windmill. One Long Island company recently began selling wind-driven units that deliver from 750 to 12,500 watts of electricity, which is enough for average home use. In the past, when the wind died down the household would have to draw on storage batteries. Integrated eco-homes, that utilize solar and wind systems and store energy, may be common in the future.

Using breezes from Lake Erie, Sandusky, Ohio, has a 100-foot experimental steel windmill. The two 62-foot aluminum blades are constructed with technology borrowed from the helicopter. Each weighs one ton and is "feathered" to turn on its axis and whirl at 40 rpm. The output, 100 kilowatts of electricity, is enough for 30 homes.

At the Sandia Laboratories in New Mexico, a large eggbeater-shaped rotor is being tested. The symmetrical shape offers the same-sized surface area to winds from any direction.[2]

Maximum storage efficiency is equally important as good blade and rotor design. For instance, on windy days, water can be pumped into reservoirs. When there is no wind, the water cascades down through hydraulic turbines. Flywheels spin at high speeds when the windmill is working and are used to run electric generators when the wind ebbs. Storage

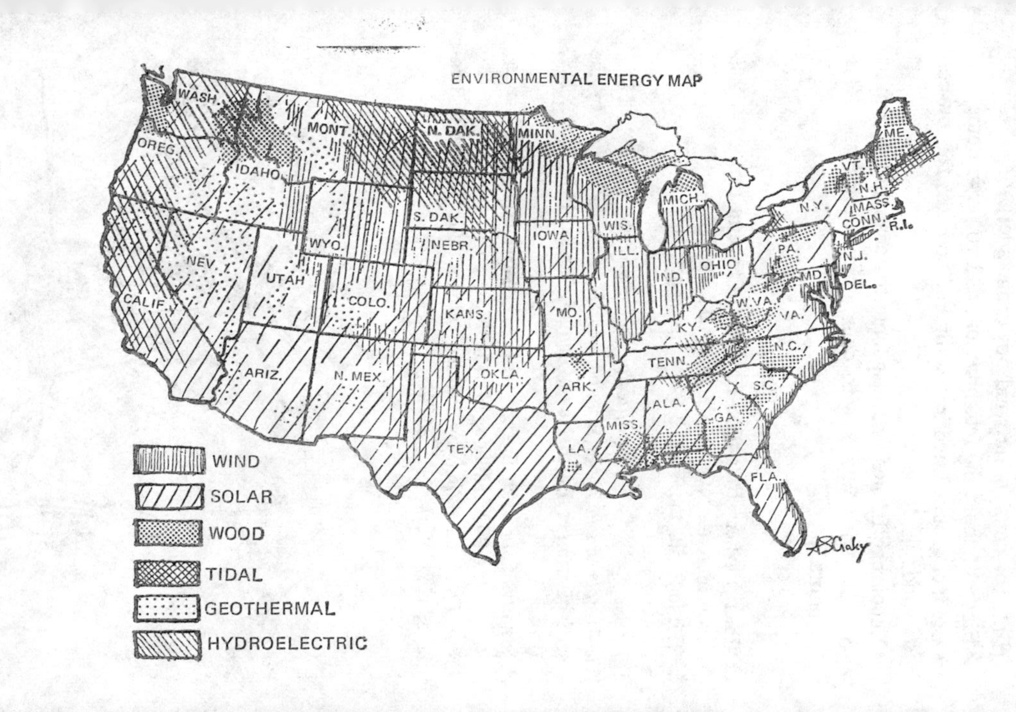
ENVIRONMENTAL ENERGY MAP
WASH.
OREG.
CALIF.
NEV.
IDAHO
MONT.
WYO.
UTAH
ARIZ.
COLO.
N. MEX.
N. DAK.
S. DAK.
NEBR.
KANS.
OKLA.
TEX.
MINN.
IOWA
MO.
ARK.
LA.
WIS.
ILL.
MICH.
IND.
OHIO
KY.
TENN.
MISS.
ALA.
GA.
FLA.
S.C.
N.C.
VA.
W.VA.
MD.
DEL.
PA.
N.J.
N.Y.
CONN.
R.I.
MASS.
VT.
N.H.
ME.
WIND
SOLAR
WOOD
TIDAL
GEOTHERMAL
HYDROELECTRIC

in batteries is not feasible when there is a heavy electrical demand during the windless period.

Hints for considering wind power include:

- Find out about the feasibility of wind power in home.
- Contact groups listed in "Additional Sources" for low-cost windmill designs.
- Encourage friends and neighbors to repair old windmills.
- Discuss wind power with community groups and legislators.

Notes

1. "Energy Alternatives–Eco-Tips, #7," Concern, Inc., 2233 Wisconsin Ave., NW, Washington, DC 20007.
2. "Tilting with the Wind," *Time* magazine, July 7, 1975.

Additional Sources

The Mother Earth News Handbook of Homemade Power, by the staff of Mother Earth News, Bantam Books, New York, NY, 1974.

Alternative Sources of Energy, #14, Alternative Sources of Energy, Rt. 2, Box 90A, Milaco, MN 56353.

Power From the Wind, by Palmer Putnam, Van Nostrand Reinhold, New York, NY, 1974.

Producing Your Own Power, edited by Carol Hupping Stoner, Vintage/Random House, New York, NY.

New Low Cost Sources of Energy for the Home, by Peter Clegg, Garden Way Publishing: Charlotte, VT 05445.

American Wind Energy Association
c/o Environmental Energies, Inc.
21243 Grand River
Detroit, MI 48219

Windpower Digest
54468 CR 31
Bristol, IN 46507

Sun Times
Alternative Energy Resource Organization
417 Stapleton Bldg.
Billings, MT 59101

II OTHER HOME CONSERVING

#15 LIGHT HOUSE EFFICIENTLY

Lighting accounts for 4 per cent of the average home fuel bill. However, in many residences and office buildings this percentage is far higher. The simplest conservation rule for saving energy is to turn lights off when not in use. When fluorescent lighting is used for an hour or so it is better both for saving energy and for life of the tube to keep lights burning. For incandescent lighting *always* turn lights off even for short periods of time.

Another fundamental rule is not to overlight the home. Most homes in this country use too many lamps and lighting fixtures which emit far too much light. The first decision for efficient and economical home lighting is to decide on the type lighting to be used: incandescent or fluorescent.

CHOOSING A LIGHT

	Incandescent	*Fluorescent*
Energy Use	Less efficient, produces more heat than light: 95% heat, 5% light.	Over 4 times as efficient as incandescent and lasts 7–10 times longer: 30% light, 70% heat.
Cost	Initial costs are less than fluorescent lighting.	Operating costs are less.
Versatility	Light can be adjusted by interchanging differ-	Usually are stationary, ceiling-type fixtures

	Incandescent	*Fluorescent*
	ent wattage bulbs. Gain flexibility by using three-way bulbs and multiple switch controls.	that provide diffused lighting.
Best Use	For specific, close-in lighting: study, sewing room, bedroom, living room, or any area that requires light for reading and writing.	For general lighting: large work areas, bathrooms, laundry room, basement, kitchen, hallways, stairways, recreation rooms, or any large room where diffused light is required.

WAYS TO SAVE ENERGY IN LIGHTING

- Turn off lights when rooms are unoccupied.
- Open curtains and shades to utilize daylight and reduce electric lighting.
- Burn candles at dinner and at parties to save energy and add to the festivity.
- Use lower-wattage bulbs.
- Use localized lighting for close work instead of lighting an entire room.
- Use a dimmer switch to allow the right amount of light for the task involved.
- Avoid colored bulbs; they are almost 60 per cent less efficient than white bulbs.
- Use light-colored lampshades; they allow maximum light into the room while reducing glare from the bulb.
- Paint with light colors; they reflect light and reduce the amount of needed lighting.
- Do not use decorative gas lamps. Four million of them in this country burn 18,000 cubic feet of natural gas yearly.[1]
- Minimize outside lighting and use low-wattage night lights.

- Reduce the use of Christmas lights and decorate for the holidays in energy-conscious ways.

ENERGY INVENTORY

See how much energy can be saved through a simple test. For one day, pursue normal activities and check the electric meter at the end of the day. The next day be conservation-minded and minimize lighting needs by keeping the switches turned off. Compare the difference in kilowatt hours between the two readings. Take the difference between the two days and multiply by 365 to find the number of kilowatt hours that could be saved in a year.

HOW TO READ YOUR ELECTRIC METER

The meter is usually located on the outside wall of a house or in the basement. The dials are read from left to right and turn in the direction of the arrows above the dial. Most meters have four calibrated dials which start from the right dial, in tens of kilowatt hours, and progress to the far left dial, which measures tens of thousands of kilowatt hours. See the diagram below for a correct reading. When the pointer is between two numbers, read the lower number.

READING: 79400

Note

1. "Turning Off Gas Flames," by Joseph C. Davis, the Washington *Post,* July 12, 1975.

Other References

Energy Misers Manual, by William H. Morrell, The Grist Mill, Eliot, ME, 1975.

"Planning Your Home Lighting," Department of Agriculture Booklet, Superintendent of Documents, U.S.G.P.O., Washington, DC 20402.

"Tips for Energy Savers In and Around the Home," Federal Energy Administration Booklet, Superintendent of Documents, U.S.G.P.O., Washington, DC 20402.

Additional Sources

How To Be Your Own Home Electrician, by George Daniels, Popular Science Publishing Co., New York, NY, 1965.

"Tips for Energy Savers," Consumer Information Department of Federal Energy Administration, Pueblo, CO 81009.

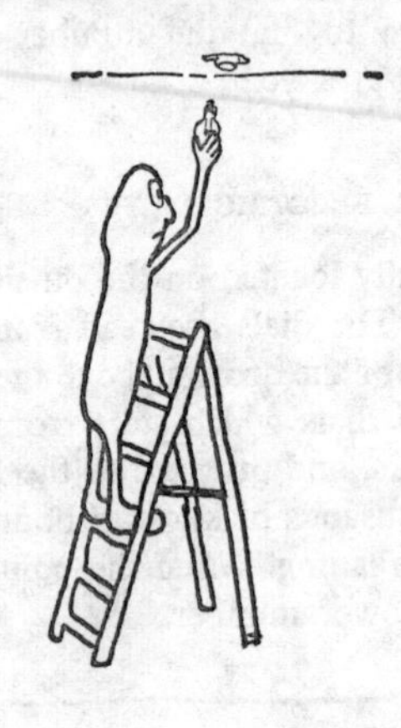

#16 SAVE WATER

As a valuable resource, clean water should be conserved. The increasing amount of waste water in our nation's sewage systems has caused public officials to look closely at water conservation efforts. One third of our waste water comes from residences while much of the remainder comes from industry. In 1970 the energy cost of treating America's waste water was 29 billion kilowatt hours or 1.8 per cent of the total electricity used.[1] After treatment, which is often inadequate and sometimes non-existent, waste water and its pollutants are dumped into rivers, streams, and coastal estuaries. Eventually, these harmful substances find their way into drinking water supply systems downstream.

Most municipal sewage systems are inadequately designed and overburdened. In most cities and towns, rain, household water (toilet and "gray" water), and industrial wastes enter the same collection system. Clean rainwater becomes contaminated by polluted water. Overload from heavy precipitation floods sewage treatment facilities and allows polluted water to pass directly into rivers and streams.

One way to reduce this waste water volume is to start a water conservation program in the home. Billions of gallons of water are wasted annually by careless homeowners. Two areas of focus are the toilet and the shower. Each time a toilet is flushed, 5 to 8 gallons of fresh water sweep away less than a pint of waste into the sewage system. Forty per cent of America's municipal water supply is used to flush toilets. Toilets can be easily adjusted to require only 3.5 gallons per flush. One simple method is to fill glass bottles with water and place upright in the tank.

A five-minute shower uses over 50 gallons of water, but a person could take a shower using half of this amount. While showering, rinse and then turn off the water while lathering; rinse afterward. Showerhead accessories may reduce the water flow from 10 gallons per minute to 3.5 gallons per minute.

As technology has grown, so has the number of gadgets that make life convenient but often more wasteful. The garbage disposal is one such invention that wastes water and energy. The nutrients in organic waste should be utilized in the garden via the compost pile, not flushed into our municipal waterways. Disposals merely chop solid garbage, mix it with water, and send it to a treatment plant. Money and energy is then spent to remove the garbage.

OTHER WAYS TO SAVE WATER IN AND AROUND THE HOME

- Repair dripping faucets; leaks can lose up to 400 gallons of water a day.
- Do not leave water running needlessly; turn off water while brushing teeth or shaving.

- Install an aerator on the kitchen faucet. This drastically reduces the volume of water flowing out, but it is hardly noticeable.
- Operate clothes washers and dishwashers with a full load.
- Save laundry rinse water to put on the garden, in compost, or to wash the car.
- In warm weather, wash the car during a rain—if there is no lightning.
- Let the rain water the garden. In a dry spell, give plants one soaking per week in the evening.
- Collect rain in a barrel or cistern for use around the house.
- Reduce flushing toilets whenever possible.

Note

1. *Energy Implications of Several Environmental Quality Strategies,* by Eric Hirst, National Technical Information Service, 5285 Port Royal Rd., Springfield, VA.

Other References

"A Primer on Water Quality," U. S. Department of the Interior Geological Survey, Superintendent of Documents, U.S.G.P.O., Washington, DC 20402.

Additional Sources

The Energy Crisis: What Can We Do? Energy Conservation Research, 9 Birch Rd., Malvern, PA 19355.

#17 AVOID AEROSOL SPRAYS

The aerosol industry was unknown prior to World War II. However, store shelves are now filled with aerosol cans containing thousands of "essential" products from air fresheners to cooking oil. Through advertising, consumers have been persuaded to buy extremely harmful, uneconomical, and wasteful products.

Aerosol sprays have become a major source of air pollution in the home. A study made by Du Pont Laboratories measured the amount of Freon propellant (low-boiling fluorocarbon liquids which expel active ingredients when volatized) suspended in air in a room ventilated with an exhaust fan. They found concentrations of HHP (Halogenated Hydrocarbon Propellants) as high as 380 and 460 parts per million in these ventilated rooms. Such concentrations are cause for concern.

Aerosol sprays have also been implicated in the development of lung cancer. Dr. William Good of Montrose, Colorado has followed cell changes in the lung with the PAP smear technique. In a study of 200 people, with one thing in common—being heavy users of aerosol sprays—he found precancerous lung cell changes in all cases.[1] However, all changes were reversed to normal after aerosol use was discontinued.

Aerosol propellants can also damage the body surface. Because they dissolve fats and remove natural oils, HHP's cause skin irritation. Spraying into the eyes can cause corneal irritation and freeze burns.[2]

In addition to Freon, other propellants used in aerosol sprays and pressurized foams can be dangerous. Nitrous oxide, used in certain food products and shaving lathers, and methylene chloride, used in hair sprays, produce an anesthetic effect. Propane, a flammable material, is also used in certain hair sprays.

Aerosol mists can produce ill effects with normal use, but

abusers can suffer tragic results. People trying to get "high" by inhaling mists are losing their lives instead. In 1972 there were over 200 such tragedies in the United States alone.[3] Since then, many more have occurred, but the exact number has not been tallied.

Hair sprays are among the most dangerous aerosols and are widely used in confined surroundings. The mist, emitted near the face, heavily contaminates the air. The ingredients have toxic effects, especially on the lungs and respiratory tract. Plasticizers in hair sprays also can be harmful. Silicone is a solvent that can damage the eyes by irritating the cornea since natural fluids cannot wash it away.

Aerosol containers can inflict serious bodily harm. Aerosol cans are marketed in metal containers with ingredients pressurized between 40 and 100 pounds per square inch. Aerosol mixtures are not dangerous under normal conditions, but over long periods of time and in high temperatures, the contents can reach bursting pressures of 210 to 400 pounds per square inch. Temperatures ranging from 120° to 160° F. are sufficient to burst the container, sending shrapnel in all directions. Furnaces, stoves, ovens, and sunlight have been known to induce aerosol explosions.

The aerosol boom also has affected the environment. The fluorocarbon Freon, a propellant used in half of all aerosols, is believed to alter the upper atmospheric ozone layer. If the ozone layer is reduced, ultraviolet rays will more easily reach the earth which could increase skin cancer cases around the world. Another environmental consequence is the disposal problem. Aerosol "bombs" cannot be recycled because they ruin metal shredders; this represents a tremendous loss of metallic resources. A 6-ounce aerosol can of cooking oil uses 2,700 Btu's to make the container and 12,300 Btu's for the contents, compared to 8,500 Btu's for the same quantity in a bottle.[4]

Consumers may wonder what to use instead of aerosols. The proliferation of aerosols on store shelves creates the impression that we cannot function without them. However, careful shopping can reveal satisfactory replacements that will save money. The net weight of an aerosol includes the weight of the propellant, which accounts for almost 70 per cent of the average aerosol product.

Using aerosols is no more effective than pouring, wiping, or dusting; it is often wasteful because the sprayed substance goes beyond the target area. When it is desirable to evenly cover a surface or to spray hard-to-reach places, use a mechanical spray gun. They are available in various sizes at hardware or department stores.

There are no substitutes for unnecessary aerosol products since these were created simply to exploit the aerosol's popularity. Products such as cooking pan coatings are especially undesirable because they are commonly used by "sniffers" to get high. Using melted fats or oil has always been satisfactory.

Pan preparation and food-flavoring sprays are excessively expensive. For example, a 9-ounce aerosol oil and butter spray can cost $1.09, while a pound of margarine costs as little as $.59. Most of the spray is propellant.

AEROSOL REPLACEMENTS

Personal Items and Toiletries

- Breath fresheners: a warm salt water rinse is better.
- Deodorants: roll-ons, creams, and sticks last longer.
- Depilatories: a razor won't cause the dermatitis that some chemicals can.
- Feminine deodorants: don't use in any form; soap and water will keep you clean.
- Hair care: use wave-setting lotions and liquid hair conditioners.

- Perfume and colognes: liquid and pump sprays will suffice.
- Shaving cream: brush and shaving soaps are more economical and safer.

Household Cleaners and Supplies

- Air fresheners: open a window, burn incense or scented candles, or leave a bowl of diluted ammonia to remove a bad odor.
- Drain cleaners: use a liquid or crystal.
- Furniture polish and wax: liquid or paste applied with a cloth is better.
- Glass cleaners: ammonia water or manual spray cleaners work just as well.
- Household cleaners: use a liquid.
- Cooking oil: oil and margarine are less costly.
- Oven cleaners: use pastes or leave a small bowl of ammonia overnight in the oven before cleaning.
- Rug shampoos: liquids or powders are available.
- Spot removers: use liquids or paste.

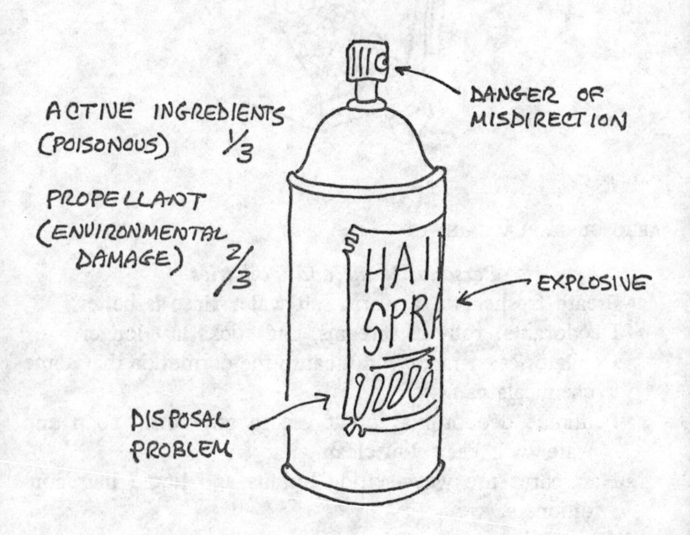

- Medical supplies: ointments, creams, and liquids are just as effective.
- Whipped creams and cheeses: try *real* food that's not processed and adulterated.

Notes

1. *Aerosol Sprays,* by Barbara Hogan and Dennis Darcey, CSPI Publications, 1757 S St., NW, Washington, DC 20009, 1976.
2. Ibid.
3. "FDA to Act on Possible Risk in Aerosol Sprays; New Rules Would Affect a Variety of Products," by J. Spivak, *Wall Street Journal,* August 11, 1972, p. 24.
4. *Energy and Food,* by Albert J. Fritsch, Linda Dujack, and Douglas Jimerson, CSPI Publications, 1757 S St., NW, Washington, DC 20009, 1975, p. 70.

Additional Sources

A Study of Indoor Air Quality, by William A. Cole, Willard A. Wade, and John E. Yocom, prepared for EPA Office of Research and Development, U. S. Environmental Protection Agency, Washington, DC, September 1974.

"The Sniffing Spectrum," *Do It Now,* PO Box 5115, Phoenix, AZ 85010.

#18 USE FEW CLEANING PRODUCTS

To sell their dazzling array of home-care products, manufacturers exploit those who take pride in housekeeping. Advertising has created a fetish against spots, stains, odors, and germs. One can buy special products to clean any area or utensil in the kitchen: walls, tiles, floors, counter tops, stoves, ovens, copper, ceramic or stainless steel pots, windows, and appliances. Most of these specialized cleaners are overpriced and only partially effective. Some, such as aerosols and phosphate detergents, have serious environmental effects. The energy used to manufacture these products and their containers is extremely wasteful. By purchasing simple cleaners one can clean the house thoroughly and inexpensively while saving energy. Money is saved because fewer products are purchased. The high cost of manufactured products often results from selling points such as sweet scents, coloring, and overpackaging.

KEEP A CLEAN HOUSE WITH SIMPLE CLEANING PRODUCTS

- Do not use aerosols!
- Powdered laundry soap: non-polluting varieties can be used to wash clothes and dishes.
- Bleach: brightens laundry, cleans, and disinfects any surface.
- Ammonia water: an all-purpose cleaner, especially good for windows; a cupful left overnight in the oven will loosen baked-on grease which can be easily removed with baking soda.
- Steel wool: the best scouring material for washing floors, pots, pans, and ovens.
- Borax: a good grease cutter that can be used anywhere in the house.
- Baking soda: a good cleaner that also deodorizes, especially in the refrigerator.
- Scouring powder: cleans kitchen and bathroom surfaces.
- Witch hazel: can be an effective deodorant if used with cornstarch.
- Soap and water: the simplest cleaner around.
- Salt: damp salt can be used as a scouring powder for cutting boards, pots, and pans.

References

The Formula Manual, Stark Research Corp., Cedarburg, WI 53012, 1974.

How To Clean Everything, by Alma Chestnut Moore, Simon and Schuster, New York, NY, 1961.

"Plain Salt: The Cheapest Cure for 15 Household Chores," by Lucille Goodyear, *Consumer Gazette,* December/January, 1975, p. 66.

Additional Sources

Soap, by Ann Sela Branson, Workman Publishing Co., New York, NY, 1975.

#19 BUDGET RESOURCES

In the 1970s, few wage earners can keep pace with double-digit inflation. Twenty-five years ago the median family income was around $3,000.[1] Today, it has increased to over $13,000, but inflation and rising taxes have combined to decrease buying power. An indication of our economic problems can be seen in a comparison of bankruptcy cases. In 1940 there were 52,320 cases, in 1974 there were 189,513 cases—over three times as many.[2]

Living a simple lifestyle can conserve family funds for important activities and purchases. Keeping exact accounts of money transactions will show how much is frittered away on non-essentials. Keep a record book, save receipts, and keep monthly accounts of how much is spent in different areas: household supplies, food, utilities, transportation, insurance, medical care, recreation, charity, clothing, cosmetics, gifts, etc.

The grocery bill is the best place to save. For example, it is not unusual for a family to spend up to $20 a month for paper products alone. Once it is clear where the money is going, a plan can be made to cut expenditures. A complete record of family spending is also helpful for income tax purposes.

Inherent in simple lifestyle is a dedication to reduce consumption. When shopping buy only necessary items. Don't be fooled by advertising gimmicks. Consuming less will save money for more essential expenditures.

Budgeting resources extends beyond monetary matters and into the broader context of energy consumption. Keeping track of energy use is not easy. CSPI's "Lifestyle Index" is a comprehensive attempt to examine an individual's energy consumption in all phases of living: household–heating, cooling, and lighting; foodstuffs–production, freight, processing and packaging, wholesaling and retailing, and home preparation; consumer products and leisure activities; transportation; and social and governmental services. Hopefully, the following inventory will help expose wastefulness and lead to a conscientious energy budget. To compute a household energy budget tally up home appliances, home lighting, cooling and ventilation, and space-heating energy usage. Use the following charts as guides.

HOUSEHOLD ENERGY EXPENDITURES[3,4]

1) HOME APPLIANCES:

The energy values* listed are average annual use per item. Multiply by the number of items in the home.

Electric Appliances:

clock	[6]	________
floor polisher	[6]	________
sewing machine	[4]	________
vacuum cleaner	[17]	________
air cleaner	[80]	________
bed covering	[54]	________
dehumidifier	[128]	________
heating pad	[4]	________
humidifier	[60]	________
germicidal lamp	[52]	________
hair dryer	[5]	________
heat lamp (infrared)	[5]	________
shaver	[0.7]	________
toothbrush	[0.2]	________
vibrator	[0.7]	________

* The Energy Unit (E.U.) as defined in the *Lifestyle Index* is equivalent to one ten-thousandth of the energy expended by the average American in 1972, or around 10 kilowatt hours.

clothes dryer	[365]	________
iron (hand)	[53]	________
washing machine (automatic)	[38]	________
washing machine (non-automatic)	[28]	________
water heater (standard)	[1555]	________
water heater (quick recovery)	[1770]	________
Gas Appliances:		
gas clothes dryer	[277]	________
gas water heater	[1170]	________
Electronic Appliances:		
radio	[31]	________
radio/record player	[40]	________
television (black & white)	[133]	________
television (color)	[185]	________
stereo	[14]	________
Preparing and Preserving Food:		
blender	[6]	________
broiler	[37]	________
carving knife	[3]	________
coffee maker	[39]	________
deep fryer	[30]	________
dishwasher	[133]	________
egg cooker	[5]	________
frying pan	[68]	________
hot plate	[33]	________
mixer	[4]	________
oven, microwave	[110]	________
range	[432]	________
roaster	[75]	________
sandwich grill	[12]	________
toaster	[14]	________
trash compacter	[18]	________
waffle iron	[8]	________
waste disposer	[11]	________
freezer	[440]	________
freezer (frostless 15 cu. ft.)	[648]	________
refrigerator (12 cu. ft.)	[268]	________
refrigerator (frostless, 12 cu. ft.)	[448]	________
refrigerator-freezer (14 cu. ft.)	[418]	________
(frostless 14 cu. ft.)	[673]	________

Gas Appliances:

refrigerator	[509]	________
cooking		
single unit	[389]	________
apartment	[350]	________
outdoor gas grill	[100]	________
	Subtotal	________ E.U.

2) HOME LIGHTING:

Average electric use for lighting in the home is [268 E.U.]. This is equivalent to burning (5) 100 watt bulbs for four hours per day.

ornamental gas lights	[668]	________
average annual use	[268]	________
	Subtotal	________ E.U.

3) COOLING AND VENTILATION:

fan (attic)	[107]	________
fan (circulating)	[16]	________
fan (rollaway)	[51]	________
fan (window)	[58]	________
electric air conditioner (room)	[510]	________
gas air conditioner	[1046]	________

Central Air Conditioning:

New England	[755]	________
Mid-Atlantic	[957]	________
East North Central	[905]	________
West North Central	[905]	________
South Atlantic	[1510]	________
East South Central	[1560]	________
West South Central	[1710]	________
Mountain	[1058]	________
Pacific	[1210]	________
	Subtotal	________ E.U.

4) SPACE HEATING:

Use the following chart for approximate estimate of space heating energy expenditure:

	Electric	*Nat. Gas*	*Oil*	*Oil/ Solar*
Northeast	6,480	5,360	6,380	
Middle Atlantic	5,800	4,800	5,720	2,210
East North Central	6,030	4,900	5,940	2,290
West North Central	5,350	4,440	5,280	2,040
South Atlantic	4,460	3,700	4,400	1,700
East South Central	4,040	3,330	3,960	790
West South Central	2,900	2,400	2,860	570
Mountain	4,910	4,060	4,840	1,870
Pacific	3,800	3,140	3,740	1,440

Subtotal ________ E.U.

Part 1 Subtotal ________ E.U.
Part 2 Subtotal ________ E.U.
Part 3 Subtotal ________ E.U.
Part 4 Subtotal ________ E.U.

Total
Per Household ________ E.U.

Divide by number of users in home
Total
Per Person ________ E.U.

Notes

1. *The 1976 U.S. Factbook,* Grosset and Dunlap, New York, NY, p. 391.
2. Ibid., p. 508.
3. *Lifestyle Index,* by Albert Fritsch and Barry Castleman, CSPI Publications, 1757 S St., NW, Washington, DC 20009, pp. 10–14.
4. *The Contrasumers: A Citizen's Guide to Resource Conservation,* by Albert Fritsch, Praeger Publishers, Inc., New York, NY, 1974.

Additional Sources

Consumer Reports, Consumers Union, Mount Vernon, NY 10550.
Consumers Research Magazine, Washington, NJ 07882.

Accounting for Everyday Profit, by J. K. Lasser, Simon and Schuster, New York, NY, 1970.
How to Live Cheap but Good, by Martin Poriss, American Heritage Press, New York, NY, 1971.
Champagne Living on a Beer Budget, by Mike and Marilyn Ferguson, Putnam Press, New York, NY, 1968.
How to Live on Nothing, by Joan Ransom Shortney, Doubleday & Co., Garden City, NY, 1961.

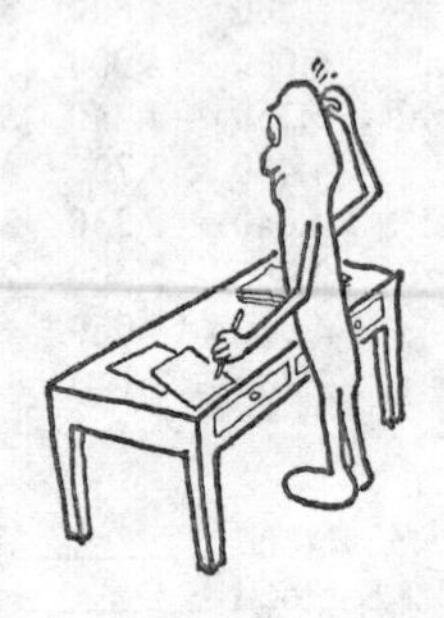

#20 BUILD A YURT

New developments are occurring around the world in simple but functional designs for shelter. William S. Coperthwaite's yurt is one that harmonizes living space with nature. It is a structure that dominates neither the landscape nor people, can be built quickly by a group of unskilled workers, and is aesthetically pleasing and inexpensive. A number of other structures have similar features, but brevity forces us to concentrate on one.

Building a yurt is a way for people to apply their skills to create family shelter. Through participating, a person can better understand and appreciate the environment. Constructing a yurt lets a person see a building's total creation in three or four days' time. Speed of construction can be important, thus providing a chance for people to feel a sense of accomplishment before becoming discouraged and abandoning the task. The yurt's simplicity of design also furnishes permanent shelter with a minimum of resources and energy.

The roots of modern yurt design lie in ancient Mongolia, where nomads of the high steppes developed a circular dwelling made of light poles, covered with skins, and supported by a tension band (similar to the hoop that holds together a wooden barrel). This protected them from the cold, violent winds of inner Asia. The permanent yurt is a modern adaptation of this ancient structure, drawing on the folk genius and simplicity of these people.

The contemporary yurt is a permanent structure made of wood and glass, using a steel cable for its support. The outward sloping walls and low eaves help the yurt to snuggle into the landscape. The outer appearance is deceptive; it seems to be a tiny space, one perhaps designed for elves or leprechauns. Within the structure, however, the reverse is true. A cascade of light falls from the central skylight. The outward slope of the walls is enhanced by the radiating lines of the roof; they meet at a ring of peripheral windows, which gives a feeling of great space. It is the quality of this inner space that makes the yurt a special place.[1] The inner space of the yurt can be easily heated with a fireplace or wood stove placed in the center of the room.

Although yurts have been used for seminar rooms, student housing, craft workshops, and mountain shelters, there is increasing interest in their use as permanent homes. The first American version was built on a campus at the Harvard Graduate School of Education. Since then, they have sprung up across the country.

THE STANDARD YURT—17 ft. (eaves) diameter

The basic design is one room. The yurt serves well as a seminar room for up to fifteen people, as student living quarters, as a summer camp, a mountain retreat, or a private office. Material costs vary from $500 to $800, depending on local prices. The plan costs $3.50.

THE CONCENTRIC YURT—32 ft. (eaves) diameter

Actually this is one yurt within another. The inner yurt supports the roof of the outer one, which saves material costs and resources. This developed from the need for a larger amount of space under one roof. This concentric way of dividing the area gives a delightful flowing space in the outer ring while providing shelter and seclusion in the inner yurt. The inner yurt is raised half a story; the resulting under-story can be used for pantry storage or can contain a couch, bookshelves, or bunks. Concentric yurts are currently in use from Maine to Alaska as permanent homes, summer homes, community centers, and library spaces. It has almost five times the room of a standard yurt and can make an excellent day-care center. With 26 windows, each 32 by 34 inches, it has abundant light. Prices of materials vary from $2,000 to $3,000. The plan costs $5.00 and is used in conjunction with the standard plan. PLANS FOR CONSTRUCTING THE YURT AND ADDITIONAL INFORMATION CAN BE OBTAINED FROM: William S. Coperthwaite, The Yurt Foundation, Bucks Harbor, ME 04618.

Note

1. Taken from *Yurt Plans* with permission of William S. Coperthwaite.

Additional Sources

Build a Yurt: The Low-Cost Mongolian Round House, by Len Charney, Macmillan and Co., New York, NY, 1974.

Your Engineered House, by Rex Roberts, available from Mother Earth Bookshelf, Hendersonville, NC 28739.

How to Build Your Home in the Woods, by Bradford Angier, Hart Publishing Co., Toronto, Can., 1969.

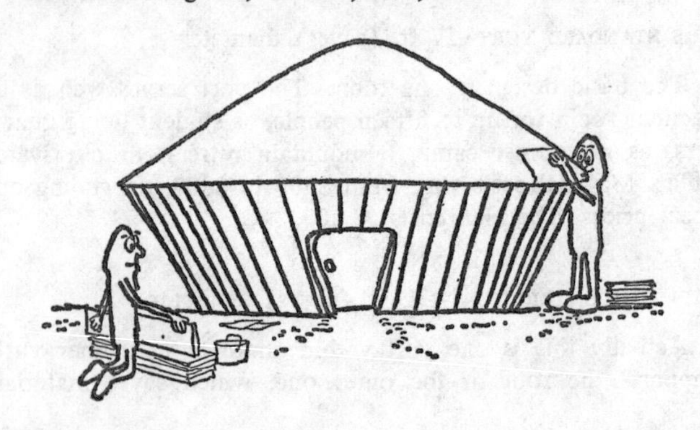

Shelter, by Dan Ljoka, Manor Books Inc., New York, NY.
How to Build a Wood Frame House, by L.O. Anderson, Dover Publishers, New York, NY, 1973.
The Owner Built Home, by Ken Kern, Charles Scribner's Sons, Totowa, NJ, 1975.
Dovebook 2, by Floyd Kahn, available from Mother Earth Bookshelf, Hendersonville, NC 28739.
Basic Construction Techniques for Houses and Small Buildings Simply Explained, prepared by the Bureau of Naval Personnel, Suitland, MD.

#21 MAKE HOME REPAIRS

Repairing simple disorders in the home can save money, resources, and energy. The following hints are elementary repair procedures which should be familiar to every homeowner.

1. *Repair a Leaking Faucet.* They can waste up to 400 gallons of water each day (see entry #16).
   Materials: • Box of assorted-size washers
   • Screwdriver
   • Adjustable wrench

How to Fix It[1]

1. First turn off the water at the shut-off valve nearest to the faucet you are going to repair. Then turn on the faucet until the water stops flowing (Fig. 1).
2. Loosen packing nut with wrench (Fig. 2). (Most nuts

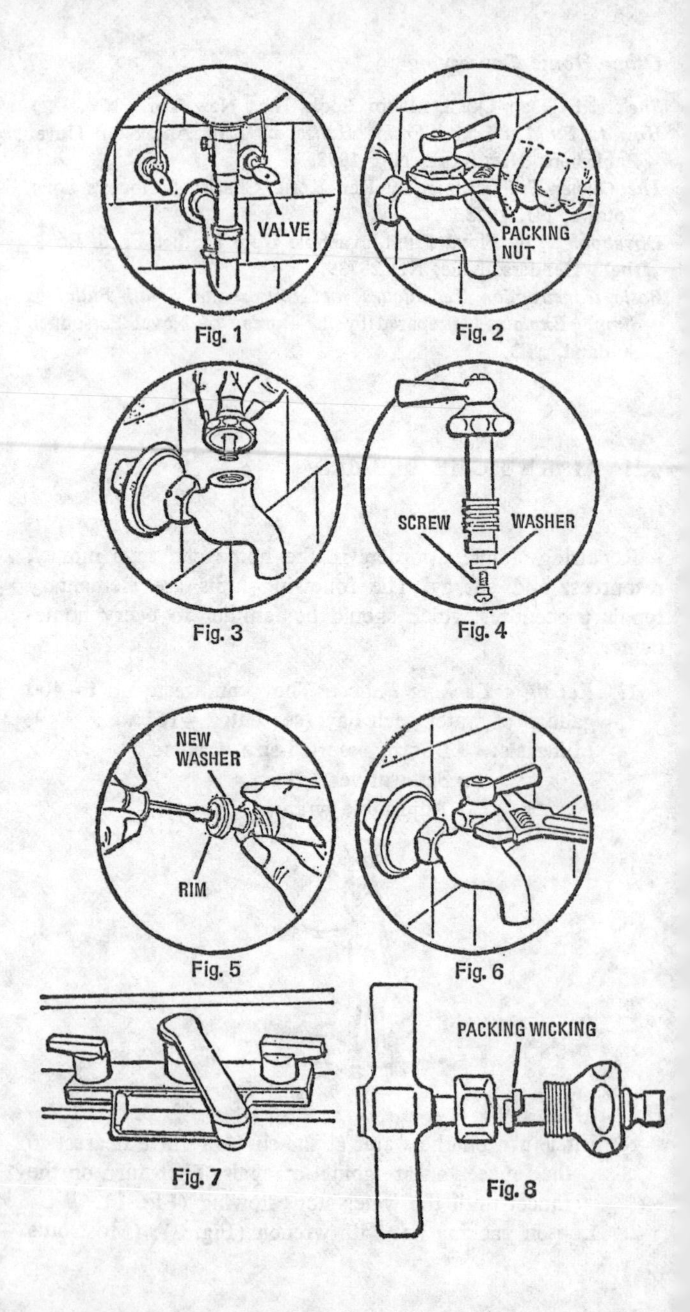

Fig. 1
Fig. 2
Fig. 3
Fig. 4
Fig. 5
Fig. 6
Fig. 7
Fig. 8

loosen by turning counterclockwise). Use the handle to pull out the valve unit (Fig. 3).

3. Remove the screw holding the old washer at the bottom of the valve unit (Fig. 4).
4. Put in new washer and replace screw (Fig. 5).
5. Put valve unit back in faucet. Turn handle to the proper position.
6. Tighten the packing nut (Fig. 6).
7. Turn on the water at the shut-off valve.

Faucets may look different, but they are all built about the same. Mixing faucets, which are used on sinks, laundry tubs, and bathtubs are actually two separate units with the same spout. Each unit will have to be repaired separately (Fig. 7).

Is water leaking around the packing nut? Try tightening the nut. If it still leaks, remove the handle and loosen the packing nut. If there is a washer under it, replace the washer. If there's no washer, you may need to wrap the spindle with "packing wicking" (Fig. 8). Then replace packing nut and handle, and turn water back on at the shut-off valve.

2. *Replace a Broken Window.* Save on heating and cooling costs with properly repaired windows.

Materials:
- Cut window glass (to correct size)
- Putty or glazing compound
- Putty knife
- Hammer
- Pliers
- Glazier points

How to Fix It[2]

1. Work from the outside of the frame (Fig. 1).
2. Remove the broken glass with pliers to avoid cuts (Fig. 2).
3. Remove old putty and glazier points. Pliers will be helpful to do this (Fig. 3).
4. Place a thin ribbon of putty in the frame (Fig. 4).
5. Place glass firmly against the putty (Fig. 5).
6. Insert glazier points. Tap in carefully to prevent breaking the glass. Points should be placed near

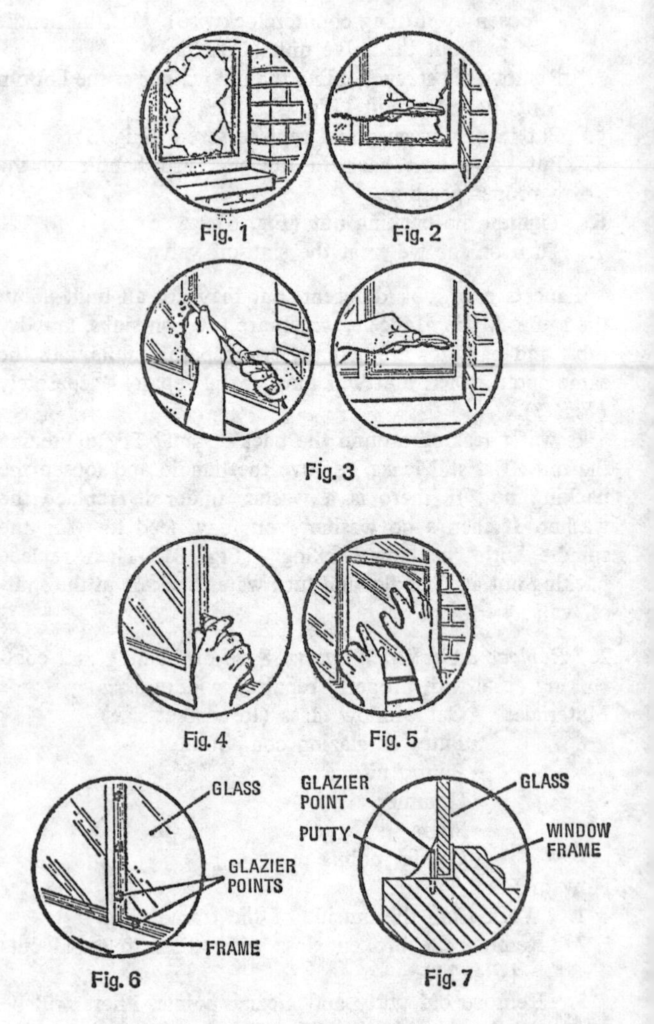

Fig. 1 Fig. 2

Fig. 3

Fig. 4 Fig. 5

Fig. 6 Fig. 7

the corners first, and then every 4 to 6 inches along the glass (Fig. 6).

7. Fill the groove with putty or glazing compound. Press it firmly against the glass with putty knife or fingers. Smooth the surface with the putty knife.

The putty should form a smooth seal around the window (Fig. 7).

3. *Repair or Replace Electric Plugs.* Faulty plugs are a safety and fire hazard. They can overload circuitry and ruin appliances.

Materials:
- Screwdriver
- Knifc
- New plug, if needed

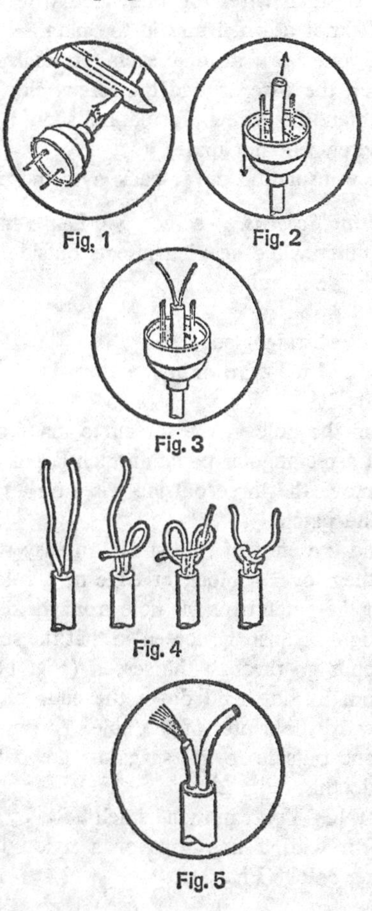

Fig. 1 Fig. 2 Fig. 3 Fig. 4 Fig. 5

How to Fix It[3]

1. Cut the cord off at the damaged part (Fig. 1).
2. Slip the plug back on the cord (Fig. 2).
3. Clip and separate the cord (Fig. 3).
4. Tie Underwriters' knot (Fig. 4).
5. Remove a half-inch of the insulation from the end of the wires. Do not cut any of the small wires (Fig. 5).
6. Twist small wires together, clockwise.
7. Pull knot down firmly in the plug.
8. Pull one wire around each terminal to the screw.
9. Wrap the wire around the screw, clockwise.
10. Tighten the screw. Insulation should come to the screw but not under it.
11. Place insulation cover back over the plug.

4. *Repairing Screens.* A screen will help ventilate a house in summer while keeping insects outside.

Materials:
- Screening
- Shears
- Straight edge
- Fine wire or nylon thread

How to Fix It[4]

1. Trim the hole in the screen to make smooth edges.
2. Cut a rectangular patch an inch larger than the hole.
3. Remove the three outside wires on all four sides of the patch.
4. Bend the ends of the wires. An easy way is to bend them over a block or edge or a ruler.
5. Put the patch over the hole from the outside. Hold it tight against the screen so that the small, bent wire ends go through the screen (Fig. 1).
6. From inside, bend down the ends of the wires toward the center of the hole. You may need someone outside to press against the patch while you do this (Fig. 2).

MENDING–You can mend small holes by stitching back and forth with a fine wire or a nylon thread. Use a matching color (Fig. 3).

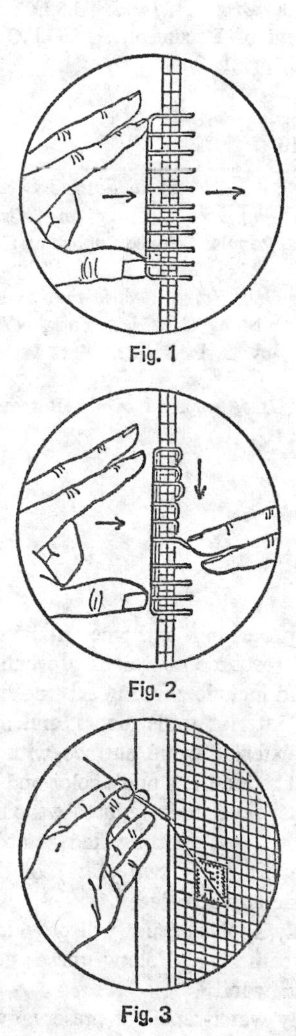

Fig. 1

Fig. 2

Fig. 3

Notes

1. *Simple Home Repairs . . . Inside,* U.S.D.A. Extension Service, Superintendent of Documents, U.S.G.P.O., Washington, DC 20402, 1973, pp. 1, 2.
2. Ibid., pp. 11, 12.
3. Ibid., pp. 3, 4.
4. Ibid., pp. 9, 10.

Additional Sources

Home Guide to Plumbing, Heating and Air Conditioning, by George Daniels, Popular Science Publishing Co., New York, NY, 1967.

How To Be Your Own Home Electrician, by George Daniels, Popular Science Publishing Co., New York, NY, 1965.

Wiring Simplified, by H. P. Richter, Park Publishers, New York, NY, 1974.

Tools and Their Uses, prepared by the Bureau of Naval Personnel, Suitland, MD.

#22 PAINT THE HOME

Adequately protecting a home from weathering is an economical and resource conscious prevention measure. This protection should include painting existing surfaces and finishing new siding, interior walls, floors, and trim.

Treating an exterior wood surface with a water-repellent preservative will retain the wood's color and retard the growth of mildew. Water-repellency reduces warping, shrinking, and swelling, which causes splitting and retards leaching. Continued use (reapply every two years) of these preservatives will prevent serious wood decay.

When properly applied, paint will provide the most protection against weathering. Follow these simple steps when painting exterior wood:[1]

1. First, apply water-repellent preservative to all joints by brushing or spraying. Treat lap and butt joints, ends and edges of lumber, and window sash and trim. Allow two warm days of drying before painting.

2. Next, prime the treated wood surface with an oil-based paint, free of lead and zinc-oxide pigment. Do not use a porous low-luster oil paint as a primer on wood surfaces. Apply enough primer to conceal the grain of the wood. Open joints should be caulked after priming.

3. Finally, apply two coats of high-quality oil, alkyd, or latex paint over the primer, especially to the side with the most severe exposure.

Interior finishes for wood, dry wall, or plaster will make the surface easy to clean, enhance the wood's natural beauty, or provide the desired color and resistance to wear. Wood surfaces can be finished with paint or a clear finish. Plaster based materials should always be painted. The interior areas and available finish materials are listed in the table.[2]

The most important part of interior painting is preparing the surface. The preliminary work is worth the trouble, for without proper preparation, one cannot obtain a lasting finish.

Begin with an inspection to find what must be done to prepare a good painting surface:[3]

- Remove old paint from woodwork. If using a paint remover, be sure there is plenty of ventilation. Cover exposed skin areas and wear goggles. Apply with an old brush. Leave until wrinkles of paint pile up and scrape with a rubber squeegee or putty knife. Apply another coat where necessary. When using a sanding machine, wear a mask over the nose and mouth. Sand with the grain of the wood.
- Remove cracked and peeling paint from walls and ceilings with a putty knife.
- Countersink nails.
- Set and putty nail holes, caulk or putty open seams, and fill nicks and cracks with spackling compound. Let dry and sand smooth.
- Scrape away rust from metal surfaces.
- Apply metal primer to nails and metal surfaces. Apply primer to surfaces not previously painted.
- Wash and clear away loose dirt, dust, oil, and wax.
- Remove mildew with a strong bleach and water solution.
- Replace crumbling putty around windowpanes and allow to dry.

Item	Interior latex	Flat oil paint	Semi-gloss paint	Floor (wood) seal	Varnish	Floor or deck enamel
Wood floors	–	–	–	X	X	X
Wood paneling and trim	–	X	X	X	X	–
Kitchen and bathroom walls (smooth surface with good scrubbability)	–	–	X	–	–	–
Dry-wall and plaster (rougher surface)	X	X	–	–	–	–

- Remove hardware, curtains, and drapes from the room. Move furniture to the center of the room, cover it, and spread newspapers on the floor.

Buy enough paint for the job. Do not use lead paint which is especially toxic to children. Do not use paint with mercury compounds indoors. Read instructions carefully. Paint during a dry spell.

Use a good brush or roller large enough for the job. First, spot paint all the repaired surfaces. Paint the ceilings first by making a large "W", then spreading to cover using an extension on the roller handle. Start walls in the upper left corner with vertical strips. Use a 2″ brush for edging and corners. Do the woodwork last (see diagram). Windows can be cov-

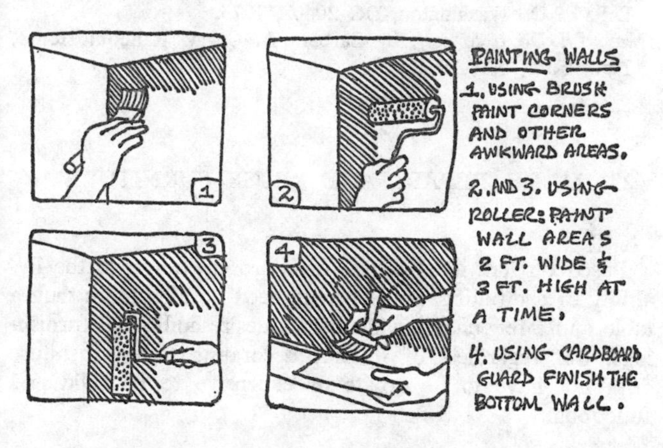

ered with masking tape, but remove the tape before the sun bakes it on.

After a day's painting is complete, make a vertical line on the outside of the can with the brush. This shows the color and amount of paint in the can. Close the can and turn it over for a few seconds to seal the lid.

Notes

1. "New Life for Old Dwellings," by Gerald E. Sherwood, U.S.D.A. Handbook #481, Superintendent of Documents, U.S.G.P.O., Washington, DC 20402, 1975, pp. 81–2.
2. Ibid., p. 85.
3. "Interior Painting," U.S.D.A. Home and Garden Bulletin #184, Superintendent of Documents, U.S.G.P.O., Washington, DC 20402.

Additional Sources

"Exterior Painting," U.S.D.A. Home and Garden Bulletin #155, Superintendent of Documents, U.S.G.P.O., Washington, DC 20402, 1973.

"Wood Siding–Installing, Finishing, Maintaining," U.S.D.A. Home and Garden Bulletin #203, Superintendent of Documents, U.S.G.P.O., Washington, DC 20402, 1973.

"Condensation Problems in Your House: Prevention and Solution," U.S.D.A. Information Bulletin #373, Superintendent of Documents, U.S.G.P.O., Washington, DC 20402, 1974.

"Principles for Protecting Wood Buildings from Decay," Forest Service, Res. Pap FPL 190, Superintendent of Documents, U.S.G.P.O., Washington, DC 20402, 1973.

Okay, I'll Do it Myself, by Barbara A. Curry, Random House, New York, NY, 1971.

#23 MAKE, REPAIR, AND REUSE FURNITURE

Based on individual tastes, furniture can improve the livability of a dwelling. Some people need nothing more than a table and a few chairs, while others desire additional furnishings. But regardless of interior decorating tastes, furniture does not have to be ornate or expensive to be solid and functional.

Families often keep furniture for years, passing it down to succeeding generations. In this way, they are assured of quality furniture with sentimental value. Check the attics or basements of relatives for furniture discoveries.

Auctions, garage sales, thrift shops, or antique stores often contain real bargains. Read the newspaper ads and check office bulletin boards for information on where to buy. Good quality used furniture is better than poor quality new items. Unfinished furniture can be bought for a fraction of the cost of a showroom piece.

When hunting for bargains choose furniture that does not require elaborate repairs. A good sanding and paint job is easy, but costly repairs will reduce the value of a bargain. Check the framework and general construction of large pieces, supports on chairs and tables, the balance and sturdiness of tables, and the pull and slide on drawers.

A little ingenuity can make something special out of something ordinary. A new slipcover will save an old sofa or padded chair, and a homemade tablecloth can conceal a badly marred table top. Denim or other sturdy cloth can make a new back on a chair.

When scratches or scars necessitate repair work, stains or antiquing kits can totally renovate a potential discard. Red wine makes a good stain; let it dry then wipe with salad oil. Antiquing kits are inexpensive and available in a variety of colors.

It is surprising to see how furnishings can be made by combining creativity with discarded furniture. All you need are sandpaper, stains, antiquing kits, cloth, boards, bricks, and a hammer and nails.

- Make small tables by stacking bricks in a circular, square, or rectangular fashion to the desired height. Use a thick board as a table top, painted the desired color and cut to the required specifications. (Bricks are heavy, so be sure the floor can support the extra weight.)
- A wooden barrel (whole or cut in half) or empty oil drum can be used as a table support. An old door or a piece of plexiglass are two other possible table tops.
- Power-line spools can serve as tables.
- Bookshelves can be made using boards and bricks.

- A bench can be made using an old chest with a long cushion.
- Individual seats can be fashioned from boxes and small cushions.

Save money and materials by making new furniture at home and repairing any old furniture. For example, to repair drawers that may be sticking, broken, or missing knobs all that is needed is a screwdriver, sandpaper, and candle.

FIX IT

For Handles and Knobs:

1. Tighten handles or knobs with the screwdriver from inside the drawer.
2. Use small thread spools to replace lost knobs.

For Sticking Drawers:

1. Remove the drawer and look for shiny places on top or bottom edges and on the sides.
2. Sand down the shiny areas and continue until the sticking stops.
3. Rub the drawer and frame where they touch, with candle wax, paraffin, or soap. This will allow the drawers to glide easier, especially when filled with heavy items. (Fig. 1.)[1]
4. If the glides are badly worn, the drawer will probably not close all the way. In this case remove it and insert

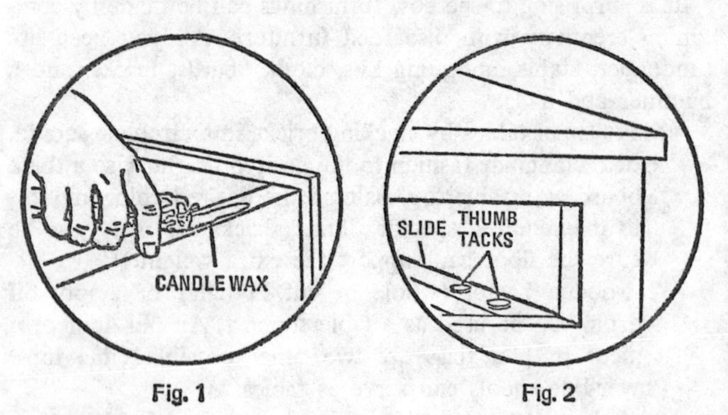

Fig. 1 Fig. 2

two or three large, smooth-head thumbtacks along the front of each glide. (Fig. 2.)[2]

Notes

1. *Simple Home Repairs . . . Inside,* U.S.D.A. Extension Service, Superintendent of Documents, U.S.G.P.O., Washington, DC 20402, 1973, p. 18.
2. Ibid., p. 18.

Additional Sources

Designer Furniture Anyone Can Make, by William E. Schremp, Simon and Schuster, New York, NY, 1972.

Furniture Finishing, Decoration and Patching, by Albert Brace Patton and Clarence Lee Vaughn, Fredrick J. Drake and Co., Chicago, IL, 1955.

Upholstery: A Complete Do It Yourself Instruction Course, by Arthur Bevin, Arc Books, New York, NY, 1962.

Cabinet Making for Beginners, by Chester Hayward, Drake Publishers, New York, NY, 1974.

Hand Woodworking Tools, by Leo P. McDonnel, Delner Publishers, Inc., Albany, NY 12205.

Fundamentals of Carpentry, by W. E. Durbahn and E. W. Surlberg, American Technical Society, 848 E. 58th St., Chicago, IL 60637, 1967.

#24 ELIMINATE UNNECESSARY APPLIANCES

The average household has around twenty-nine small electrical appliances that use almost 10 per cent of the energy consumed in the home and cost precious dollars to operate. Some of these gadgets are necessary and helpful in everyday living, but the majority are "convenience gimmicks" that are wasteful and inefficient. Simple tasks like carving meat, combing hair, brushing teeth, washing dishes, squeezing juice, and opening cans should be performed with human energy and not electrical energy.

QUIZ ON APPLIANCE GADGETRY

1. Count the number of electrical appliances in the home. ________
2. Which ones are used frequently? ________
3. Which ones are really necessary? ________
4. Which can be easily done without? ________

To help answer these questions, a chart is presented below that will help identify those gadgets that are unnecessary and waste energy.

Electric Appliances	*Kilowatt Hours Consumed Annually*[1]	*Substitute*
Blender	15	egg beater and strong arm
Carving knife	8	sharp knife
Coffee maker	106	do without
Dishwasher	363	wash dishes by hand
Egg cooker	14	cook eggs in boiling water used for tea or coffee
Garbage disposal	30	compost food waste
Trash compactor	50	manually shred trash and crush cans
Clothes dryer	993	hang clothes to dry outside in summer and inside in winter (to help humidify)
Iron	144	wear permanent-press clothing

Electric Appliances	*Kilowatt Hours Consumed Annually*[1]	*Substitute*
Electric blanket	147	sleep with extra blankets
Humidifier	163	put a pan of water on thc radiator, hang up wet clothes, or open the bathroom door after a shower
Portable heater	176	heavier clothing
Hair dryer	14	dry towel, strong arms
Germicidal lamp	141	do without
Sun lamp	16	go outside and enjoy the sunshine
Shaver	1.8	grow a beard
Toothbrush	.5	brush by hand
Vibrator	2	physical exercise
Clock	17	wind-up clock
Can opener		manual can opener
Lawn mower		hand-powered mower
Hedge clipper		hand-powered clipper
Saw		hand-powered saw
Paint mixer		hand-powered mixer

Note

1. *The Contrasumers: A Citizen's Guide to Resource Conservation,* by Albert J. Fritsch, Praeger Publishers, New York, NY, 1974, pp. 100–2.

Other References

Lifestyle Index, by Albert J. Fritsch and Barry Castleman, CSPI Publications, 1757 S St., NW, Washington, DC 20009.

"The De-Energizer," Canadian Office of Energy Conservation, Toronto, Can., March 1975.

100 Ways to Save Energy and Money in the Home, Canadian Office of Energy Conservation, Toronto, Can., 1974.

"Tips for Energy Savers In and Around the Home and on the Road," by Federal Energy Administration, Washington, DC, 1975.

Additional Sources

How to Repair Small Appliances Vol. I and II, by Jack Darr, H. W. Simms Publishers, Indianapolis, IN, 1961.

"Energy Labeling of Household Appliances," National Bureau of Standards, Room A–600, Washington, DC 20234.

Home Appliance Servicing, by Edwin P. Anderson, Audel and Co., Indianapolis, IN 46206.

"Electricity Demand: Project Independence and the Clean Air Act," by the National Science Foundation Environmental Program, Oak Ridge National Laboratory, Oak Ridge, TN, November 1975.

"Kilowatt Counter: A Consumer's Guide to Energy Concepts, Quantities and Uses," Alternative Sources of Energy, Rt. 2, Box 90A, Milaca, MN 56353.

III FOOD

> I learned from my two years' experiment that it would cost incredibly little trouble to obtain one's necessary food, even in this latitude; that many a man may use as simple a diet as the animals, and yet retain health and strength.
>
> Henry Thoreau
> *Walden*, 1854

INTRODUCTION

Food shortage is a recurring global problem not solved as yet by the advances of science and technology. Millions of people eat inadequate diets, and deaths from hunger and food-deficiency diseases are common. Four hundred million people suffer from malnutrition; half of these are children who may suffer from stunted growth or mental retardation. Some of these unfortunate people live in food-exporting countries—like the United States.

As a limited world resource, food requires land to grow and energy to irrigate, transport, process, refrigerate, and cook. As much as one quarter of agricultural produce is devoured by insects and rodents, or rendered inedible due to improper storage and handling. Large portions of produce that could be used for human consumption is fed to pets and livestock. Food is also wasted in homes and restaurants.

Both food scarcity and overabundance can be unhealthy. Food intake beyond that used for daily energy needs is converted into fat. We have abundant knowledge about minimal food needs (e.g. 2,000 calories and 50 grams of protein), but millions of people in Third World countries don't reach these levels. Most Americans, however, consume so much food that even with poor dietary choices, we obtain sufficient body-building protein and calories.

The problem is not only one of production and consumption, but also one of distribution. Food is not reaching those who need it while being overconsumed by others. One response should be to arouse the affluent to conserve and share resources. A reduction in U. S. food consumption, especially of overly processed and "empty" foods, may be beneficial from a health standpoint. We should increase exports of high-protein grain and oil-bearing seeds (soybean, peanut, cotton) to countries in need. This may require that we reduce our consumption of resource-intensive foods such as grain-fed beef (currently about 116 pounds per person per year).

From 1960 to 1972 in the U.S., the amount of grain-fed cattle increased by 55 per cent. If raised on grass, the beef would be lower in fat, and the grain could be used by humans in need. Grain-fed beef is marbled with fat, which is 33 to 50 per cent saturated.

To ensure food for all, we should decrease the amount we eat, substitute vegetable protein for animal protein, minimize food waste, and add to home-food supplies by farming and gardening on available land.

To a large extent, spiraling food prices often result from food processing and overpackaging. These costs once accounted for less than 1 per cent of the food dollar, but rising energy costs have increased processing and packaging expenses. Between 12 and 16 per cent of our country's energy budget is spent on food production and preparation. This waste of resources and energy should concern every consumer because we can no longer afford such extravagance.

Stretching the dollar and good nutrition is tied to conservation. Simple hints such as growing produce at home, eating less meat, baking bread, eating wild foods, and avoiding overly packaged and processed foods not only save money, but also constitute an affirmative step toward reducing consumption of our dwindling resources and energy.

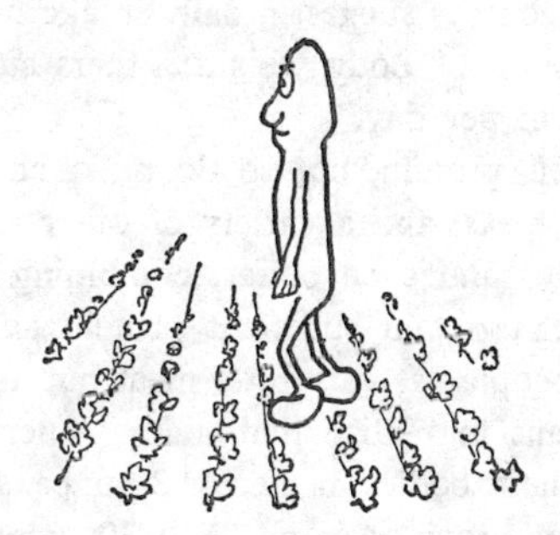

#25 CONSUME LESS MEAT

> The alternate diet is designed to prevent diseases and, at the same time, be nutritionally adequate. Because it is largely but not completely derived from legumes (pod vegetables: beans, peas, peanuts, etc.), grains, vegetable and fruit products, it is less expensive to produce in terms of resources than the present American diet based much more on food products derived from animals. It has its additional feature of ecological soundness at a time of world food shortage.
>
> Panel on Nutrition and Health of
> Senate Select Committee on Nutrition
> and Human Needs–1974

For a number of reasons, many Americans are reducing their consumption of meat. Some hold an aversion to killing animals; others want to reduce their intake of saturated fat and increase fiber in the diet; others find non-meat food sources much more economical; and still others want to share the nation's grain with the world's hungry rather than using it to fatten cattle. Whatever the rationale, a well-balanced diet can be maintained without eating meat.

The key to dietary balance is protein. The body needs a daily intake of protein for proper growth and body functions. A person's protein requirements vary with body weight and activity. Some sources suggest a daily intake of 0.36 grams of protein per pound of body weight; others suggest 50 to 60 grams of protein per day.

Meat contains protein, but so do milk, cheese, fish, eggs, nuts, soybeans, peas, and a variety of other foods. By eating meat alternatives and even better, combining them with one another, one can obtain sufficient protein and reduce meat intake. Many people eat very little meat but are very healthy.

It is important to realize that many Americans eat more protein than their bodies need. A 3-ounce serving of fish, poultry, or meat provides from 20 to 30 grams of protein; a quart of milk, 30 grams; two eggs and a slice of cheese, at least 20 grams; a cup of soybeans, 25 grams. Knowing these values, who needs a 16-ounce steak dinner?

To produce one pound of marketable meat requires:[1]

Forage-fed beef–29,650 (Btu/lb.)

Grain-fed beef–42,600 (Btu/lb.)

Pork–29,400 (Btu/lb.)

Chicken–19,150 (Btu/lb.)

Soybeans require 9,000 Btu/lb. and dried peas and beans, 13,250 Btu/lb.

In spite of this energy difference, U.S. consumption of beef has increased from 55 pounds per person in 1940 to 116 pounds per person in 1974.[2]

Soybeans are an especially good meat substitute because they contain almost all the essential amino acids, little sugar, no starch, and are high in calcium and B vitamins. The economical aspects are apparent when one considers that a 3-ounce hamburger costs 21 cents; an 8-ounce cup of soybeans costs 12 cents. Such substitutions can add up to significant savings.

Soybeans can be used a number of ways:

- Baking Bread–Replace one cup of white flour with soy flour.
- Boiled–Soak beans overnight. Place soaked beans in pot and bring to boil. Simmer 2½ hours; add ½ teaspoon pepper, salt, 1 clove minced garlic, and pork; simmer another 15 minutes.

- Baked–Prepare as in boiling; instead of simmering, bake with blackstrap molasses or diced vegetables.
- Grits–(beans broken into small pieces) Add 1 cup grits to boiling water; let fluid absorb; cool, cover, and store in refrigerator; add to omelets, souffles, stuffed peppers, or any other dish to substitute for meat.

Many people who want to begin a meatless diet have trouble getting started because they simply are not familiar with high-protein vegetarian meals. The following menu prepared by Arlene Gorelich in observance of Food Day 1976 offers nutritious and tasty eating at a low price.

Breakfast
Whole wheat-apple pancakes with maple-walnut yogurt topping
Half grapefruit
Herb tea

Lunch
Open-faced melted cheese sandwiches with tomatoes and sprouts
Lentil salad
Cantaloupe wedges
Iced mint tea

Afternoon snack
Raisins

Dinner
Eggplant-chickpea ragout with
Brown rice
Spinach salad
Steamed broccoli
Fresh-fruit salad
Herb tea or water

Evening munchies snack
Banana nut bread
Skim milk

Standard servings of these foods would provide 2,083 calories. The day's menu consists of 15% protein (78.3 grams), 57% carbohydrate (303.6 grams) and 28% fat (62.2 grams), most of which is polyunsaturated vegetable oil. Analysis of foods is based on *Food Tables of Portions Commonly Used,* Bowes and Church, 12th edition (1975), and *Home Economics Research Report #36,* U. S. Department of Agriculture, August 1969.

RECIPES

Whole Wheat-Apple Pancakes

This recipe was inspired by the pancake recipes in the Tassajara cookbook.

2 cups whole wheat flour
2 eggs
2 cups skim milk
2 medium cooking apples
1 teaspoon cinnamon
2 tablespoons vegetable oil
1 tablespoon honey
2 teaspoons baking powder
1 teaspoon salt
½ teaspoon nutmeg, mace

Grate the apples. (You should have about 1 cup grated apple.) Beat the eggs. Combine eggs, apple, milk, honey, and oil. Mix or sift together flour, salt, baking powder, and spices. Make a well in dry ingredients, add liquid ingredients, and stir. Cook pancakes on a hot griddle, lightly oiled. Serves 6.

Maple-walnut yogurt topping

To 1½ cups of plain yogurt, add 3 tablespoons of maple syrup and 3 tablespoons of crushed walnuts. Stir, and spoon onto hot pancakes.

Melted cheese sandwiches

2 slices whole wheat bread
2 ozs. cheddar cheese
Basil
1 tomato
¾ cup mung bean sprouts (or other bean sprouts)

Preheat oven to 400°. Lightly toast bread. Top bread with sliced tomatoes, basil, sprouts, then shredded cheese. Bake 5 minutes or until cheese melts.

Lentil Salad

1 cup dry lentil beans
1 small onion
1 green pepper
2 tablespoons vegetable oil
2 tablespoons wine vinegar
½ teaspoon each oregano, thyme, rosemary, tarragon

Rinse lentils thoroughly. Place in pot with 5 cups cold water.

(Add a bay leaf for fuller flavor.) Bring to boil, then simmer for 25 minutes or until lentils are tender. While lentils are cooking, mince the onions, dice the green pepper, and combine them with oil and vinegar. When lentils are cooked, drain them, and immediately add vegetable-dressing mixture. Stir, add salt to taste, and chill. May be served on lettuce. Serves 6.

The afternoon snack provides for 1 large handful of raisins, about ¼ cup.

Eggplant-chickpea Ragout

1 cup dry chickpeas (garbanzos)
1 eggplant, weighing about 1 lb.
2 lbs. tomatoes
2 onions
juice of ½ lemon
3 tablespoons vegetable oil
2 teaspoons dill weed

The night before, wash the chickpeas thoroughly; soak overnight in 5–6 cups water. To cook, put peas into 5–6 cups cold water (you can use the water they soaked in), bring to boil, and simmer about 1 hour. (In a pressure cooker, allow about 15–20 minutes.) While they cook, dice the eggplant into ½-inch cubes, slice the onions, dice the tomatoes. Heat oil in a heavy frying pan. Add dill, onions, sauté until onions are translucent. Add eggplant, cooking until eggplant browns. Add tomatoes, chickpeas, and lemon juice; salt to taste; and allow to cook 10–15 minutes more. Serves 6. Can be eaten over or alongside brown rice.

Brown Rice

Prepare like refined rice. Proportions of water to rice should be about 1½–2 cups water to each cup of raw rice. Bring to a boil, simmer about 45 minutes or until water is absorbed.

Spinach Salad

¾ lb. of raw spinach
2 chives (green onions)
1 clove garlic
¼ cup olive oil
2 tablespoons lemon juice

Wash, trim spinach; drain. Crush garlic, mince chives, add oil and lemon juice. Shake vigorously. Toss spinach with dressing. Serves 6.

Steamed Broccoli

Wash broccoli. Place on steaming rack above 1–2 inches of boiling water. Steam 15–20 minutes. If you don't have a steaming rack, put broccoli in pot with 1 scant inch of water; bring to a boil, turn down light, and simmer about 15 minutes. Broccoli is done when stalks are tender but still crisp (*al dente*). One large head–6 or 7 stalks–should feed 6.

Fruit Salad

2 navel oranges
2 large bananas
2 pears
Cinnamon

Peel, slice the oranges; slice the bananas; chunk the pears. Toss together, serve with a dash of cinnamon. No sweetener necessary!

The evening snack provides for a glass of skim milk.

Banana Nut Bread

2 cups whole wheat flour
½ cup oil
½ cup honey
2 eggs
1 teaspoon baking soda
2 cups ripe banana, mashed
½ cup chopped walnuts

Preheat oven to 350°. Beat eggs, blend honey, oil, and beaten eggs until smooth. Mix flour, baking soda, and salt. Add this dry mix to the honey batter in thirds, alternating with the banana pulp. Stir. Fold in the chopped nuts. Bake in greased loaf pan 50 minutes or until a toothpick stuck into the center of loaf comes out clean. Cool for 5 minutes, then remove from pan.

Notes

1. *Energy and Food,* CSPI Energy Series VI, CSPI Publications, 1757 S St. NW, Washington, DC 20009, 1975.
2. "National Food Situation," Economic Research Service, U. S. Department of Agriculture, Washington, DC, February, 1975.

Other References

"Breaking the Meat Habit," by Maya Pines; *Food for People, Not for Profit,* edited by Catherine Lerza and Michael Jacobson, Ballantine Books, New York, NY, 1975.

Let's Cook It Right, by Adelle Davis, Harcourt, Brace Jovanovich, Inc., New York, NY, 1970.

Diet for a Small Planet, by Frances Lappe, Ballantine Books, New York, NY, 1975.

Additional Sources

The Vegetable Passion, by Janet Barkas, Charles Scribner's Sons, New York, NY, 1975.

Meatless Cooking Celebrity Style, by Janet Barkas, Grove Press Inc., New York, NY, 1975.

The New York Times Natural Foods Cookbook, by Jean Hewitt, Garden Press, New York, NY, 1968.

The Oats, Peas, Beans & Barley Cookbook, by Edyth Young Cottrell, Woodbridge Press, Santa Barbara, CA, 1974.

Recipes for a Small Planet, by Ellen B. Ewald, Ballantine Books, New York, NY, 1973.

The Soybean Cookbook, by Mildred Lager and Dorothea Van G. Jones, Arco Publishing Co., New York, NY, 1968.

"Nutritive Value of Foods," USDA Home and Garden Bulletin No. 72, Superintendent of Documents, Washington, DC 20402.

The Supermarket Handbook: Access to Whole Foods, by Nikki Goldbeck, Harper & Row, New York, NY, 1973.

The Vegetarian Epicure, by Anna Thomas, Knopf Publishers, New York, NY, 1972.

The Deaf Smith Country Cookbook, by Marjorie Winn Ford, et al., Macmillan and Co., New York, NY, 1973.

International Vegetarian Cookery, by Sonja Richmond, Arc Books, New York, NY, 1965.

Vegetarian Gourmet Cookery, by Alan Hooker, 101 Productions, San Francisco, CA, 1970.
French Gourmet Vegetarian Cookbook, by Rosine Claire Millbrae, Celestial Arts, San Francisco, CA, 1975.
Food for Thought, by Marilyn King and William Scott, Fair Dinkum Books, Cornwall, England, 1974.

#26 SELECT UNPROCESSED FOODS

Food processing is the sixth highest fuel and electricity user in American industry, using 281.0 billion kilowatt-hours equivalent energy in 1974.[1] When fuel processing and electrical generation losses are included, the amount more than doubles. Compounding these massive energy expenditures is the increased use of energy-intensive food-packaging materials such as plastics and aluminum, which take one-quarter again as much energy as does food processing.[2]

Processing energy is wasted on TV dinners, frozen vegetables, flavorings, spreads, and canned beverages. Generally, processed and prepared foods consume much more energy than fresh foods.

Food processing technology developed in the twentieth century in order to provide urban dwellers with adequate food. Initial efforts attempted to prevent spoilage, discoloration, and dryness of fresh produce and meats. Packaging made products easier to stock and more attractive, and additives were used to flavor, preserve, thicken, or color products. Processing is so much a part of today's life that a child might think that eggs arise from styrofoam trays and milk flows from cardboard cartons.

A person's physical and mental well-being is directly related to what is eaten. Americans have such an array of available foods that we eat too much and often not wisely. Our rich and refined national diet may contribute to side effects such as obesity, heart attack, high blood pressure, hypertension, diabetes, tooth decay, and bowel cancer. A sedentary life, urbanization, and automation combine with high-calorie, low-nutrition foods to compound these problems. The irony is that people who eat whole grains, nuts, and fruits are called "food faddists," while "normal consumers" indulge in sugar-packed snacks and refined flour.

The health of Americans could be vastly improved if we followed the basic nutritional guidelines so often repeated: 1) Eat low-fat animal products; 2) Eat more fibrous foods: fruits, vegetables, whole grains; 3) Eat fewer salty, greasy, sugary snacks and junk foods.

Fibrous foods are generally rich in nutrients, low in calories, and help fill a person to prevent overeating. Roughage aids digestion and can help prevent constipation by providing bulk to facilitate passage of the stool.

Refined white flour, used in white bread, cakes, pastry, and rolls goes through a milling process that robs the grain of nutrients and fiber. Bran and wheat germ, the most nutritious part of the wheat berry, are removed. "Enriched" white flour returns some of these milled-out vitamins and minerals, but it is still inferior to whole wheat flour. When baking, use whole wheat flour.

White rice goes through a similar process when milling removes the bran and germ. Enriching will restore some of the B vitamins and iron; parboiling (converted) rice prevents some loss of nutrients during milling. However, brown rice is superior, containing all of the nutrition in the grain.

Spaghetti, macaroni, and noodles made with whole wheat flour can be obtained and are more nutritious than varieties using white flour.

In recent years, the controversy over food additives has become a major health and economic concern. Sweeteners, preservatives, coloring agents, thickeners, and extenders are added to foods. Sodium nitrite is added to cured meats to make them seem lean and tasty; artificial color is added to imitation fruit drinks to provide the appeal of real fruit; the rich yellow color of butter and egg bread is probably from yellow coloring—not butter and eggs.

Certain additives prevent spoilage and inhibit bacteria,

thereby reducing food poisoning and health problems. Others increase a food's shelf life by prolonging apparent freshness or minimizing melting properties. Not all additives are harmful—but many are unsafe. Since 1955 the number of additives has doubled, and now each American consumes about five pounds of these chemicals per year.[3]

Sodium nitrite, used as a preservative in cured meats, ham, bacon, hot dogs, and smoked fish, may cause cancer in humans. Cyclamates were banned in 1970, but now this ban is being removed. Hyperkinesis, a disorder common to young children, may be linked to coloring and flavoring agents in snack foods. That luscious, bright color of many citrus fruits may be from dye and wax—an unpalatable thought in itself.

To reduce the intake of additives, read the listed ingredients on packaging labels. If it reads like a chemistry textbook—beware. Buy fresh produce from farmers, farmer's markets, co-operatives, or grow them at home. Wash all produce; home preserve fruits and vegetables for later use. Make homemade breads, beverages, and main courses using wholesome and unprocessed ingredients. They not only will taste better, but will be more nutritious.

If possible, avoid convenience foods, non-nutritious snacks, and soft drinks. One pays for their high packaging and processing costs while gaining few nutritional benefits. For example, a 16-oz. non-returnable Royal Crown Cola bottle consumes over 6,000 Btu's for the container and half that for contents, which are not nutritious anyway.

Notes

1. "Fuels and Electric Energy Consumed," Annual Survey of Manufacturers, 1974, U. S. Department of Commerce, Washington, DC 20233, 1975.
2. *Energy and Food*, by Albert Fritsch, Linda Dujack and Douglas Jimerson, CSPI Publications, 1757 S St., NW, Washington, DC 20009.
3. *Eater's Digest–The Consumer's Factbook of Food Additives*, by Michael Jacobson, Anchor Books, Doubleday & Co., Garden City, NY, 1972.

Other References

Food for People, Not for Profit, edited by Michael Jacobson and Catherine Lerza, Ballantine Books, New York, NY, 1975.

Nutrition Scoreboard, by Michael Jacobson, Avon Books, New York, NY, 1975.

Additional Sources

The Composition of Foods, by Bernice Watt and Annabell Merrill, Superintendent of Documents, U.S.G.P.O., Washington, DC 20402.

"Cancer Prevention and the Delaney Clause," Health Research Group, 2000 P St., NW, Washington, DC 20036.

Home Bakebook of Natural Breads and Goodies, by Sandra and Bruce Sandler, Stackpole Books, Harrisburg, PA, 1972.

Cooking With Whole Grains, by Mildred E. Orton, Farrar, Strauss, and Giroux, New York, NY, 1971.

The Wonderful World of Natural Food Cookery, by Eleanor Levitt, Hearthside Press, Great Neck, NH, 1971.

A Consumer's Dictionary of Food Additives, by Ruth Winter, Crown Publishers, Inc., New York, NY, 1972.

The Complete Sprouting Cookbook, by Karen Cross Whyte, Troubadour Press, San Francisco, CA, 1973.

Yoga Natural Foods Cookbook, by Richard Hittleman, Bantam Books, New York, NY, 1970.

The Natural Foods Cookbook, by Beatrice Trum Hunter, Pyramid Publications, Inc., New York, NY 10022, 1972.

More-with-less-Cookbook, by Doris Janzen Longacre, Herald Press, Scottsdale, PA 15683, 1976.

#27 AVOID NON-NUTRITIOUS FOOD

The processed-food industry developed by exploiting the indiscriminate buyer and his/her preference for sweets. Advertising, attractive packaging, and accessibility combine to make "work saving" convenience foods and snacks quite popular. The high cost and nutritional worthlessness of these products are two reasons to eliminate them from the shopping list.

Our worsening state of national health could be related to the changes in American eating habits since 1940: Per capita consumption of dairy products has declined 19 per cent; fresh vegetables have declined 17 per cent; fresh fruits have declined 44 per cent.[1] On the other hand fats and oils increased 15 per cent, refined sugar 8 per cent, and soft drinks over 100 per cent. The pastry and fried snack industry has boomed since the Second World War.

There is nothing wrong with between-meal snacks, unless they contribute to excessive calorie intake. Problems arise when people eat too much of the wrong kind of snack foods. Sugary cakes, salty fried tidbits, and high-calorie soft drinks can contribute to diabetes, tooth decay, constipation, obesity, and hypertension.

Nutritious snacks should replace junk food in the diet. Fresh fruits and berries are refreshing anytime; dried fruits provide natural sugar to satisfy a sweet tooth; roasted nuts and soybeans provide protein. A wedge of cheese or peanut butter on whole wheat crackers will assuage a heartier appetite. Try low-calorie crunchy, raw vegetables such as carrots, celery, or cauliflower.

Soft drinks, which often cost more than milk or fruit juice, are little more than sugared carbonated water and provide no nutrition. Soft drinks and imitation fruit drinks offer nothing but empty calories and artificial colors.

Snack cakes are often advertised as nutritious since they are enriched or vitamin fortified. However, these claims are absurd, since one could get better nutrition from natural food and avoid the 300 to 500 extra calories from the sugar and fat.

Junk foods include convenience items–pot pies, frozen dinners, prepared stews, or main-course creations–that require minimal home input to prepare. Advertisements induce consumers to buy these frozen, dehydrated, artificially colored and flavored foods to avoid simple kitchen work such as boiling water or cutting vegetables. Convenience-food items are not worth the high price one must pay, nor do they compare in taste to home-prepared meals.

These quick dish foods and frozen dinners are oriented more toward quantity than quality. Starch and thickening agents provide much of the bulk. Read the listed ingredients, then put these products back on the shelf.

Don't become a prisoner of prepared meals and advertising gimmicks. Experiment with various seasonings and try new food combinations. Enjoy home-cooked meals and take pride in culinary creativity.

An additional expense of convenience items is due to excessive packaging costs. Advertising and promotion has replaced product protection in packaging materials. Throwaway bottles and aluminum cans are tremendous energy wastes and environmental blights. Shiny wraps, cardboard boxes, plastic bags, and layers of paper are virtually unnecessary and wasteful.

Up to one third of commercial and household refuse is packaging waste. Consumers should realize that they pay the high costs of excessive packaging and should curb their purchase of these items.

Note

1. *The U.S. Factbook–1975,* Grosset & Dunlap, New York, NY, p. 90.

Other References

Food for People, Not for Profit, edited by Catherine Lerza and Michael Jacobson, Ballantine Books, New York, NY, 1975, p. 165.

Nutrition Scoreboard, by Michael Jacobson, Avon Books, New York, NY, 1975.

Eater's Digest—The Consumer's Factbook of Food Additives, by Michael Jacobson, Anchor Books, Doubleday & Co., Garden City, NY, 1972.

Additional Sources

The Deaf Smith Country Cookbook, by Marjorie Winn Ford, et al., Macmillan Co., New York, NY, 1973.

The Completely Organic Directory, by Jerome Goldstein and M. C. Goldman, Rodale Press, Emmaus, PA 18049.

"CNI Weekly Report," edited by Stephen Clapp, Community Nutrition, Washington, DC.

#28 REDUCE INTAKE OF REFINED SUGAR

Everyone loves sweets. It's a preference that is often hard to resist. Ending a meal with something sweet is a common practice all over the world. With one fourth of the population overweight and modern life's conveniences reducing our physical exercise, we should reduce our sugar intake. The average American consumes about 2 pounds of refined sugar per week, representing almost 20 per cent of a person's caloric intake. In a hundred years' time, per capita U.S. sugar consumption has tripled.

Refined cane or beet sugar (sucrose) and corn sugar (dextrose and glucose) provide nothing but calories when added to food–no vitamins, minerals, or protein. The health consequences are well known to everyone–tooth decay, obesity, and diabetes. The sugar-refining process is also a heavy energy consumer.

A major reason for the increased sugar consumption is the greater availability of candy, cakes, cookies, soda, ice cream, and pre-sweetened breakfast cereals. The consumer may not be aware of how much sugar is present in many foods, such as canned fruits in heavy syrup, prepared desserts, ketchup, and even baby food (sometimes up to 50 per cent). When buying a product, read the label: if sugar is the first item listed, it is the main ingredient.

Advertisers often take advantage of the human fondness for sweets, especially in children. Exploiting the love for sweets has created a breakfast cereal industry that produces highly refined grains loaded with sugar and lacking in nutrients.

As alternatives to sugar, the main advantage of using sorghum, honey, or blackstrap molasses is that they are not highly refined; therefore trace minerals are present. Blackstrap molasses is high in iron and calcium. However, they are still a source of concentrated calories, and if a person wants to reduce caloric intake, these foods will not help one lose weight.

Natural sugar can also be obtained in dried fruits–raisins, dates, prunes, and especially apricots. Dried fruits will provide vitamins and minerals and are better energy sources than refined sugar.

Instead of syrup and jelly on pancakes, waffles, or toast, try using berries or fruits. For dessert, serve fruits and nuts instead of rich cakes, ice cream, or pies. Drink natural fruit juices instead of soft drinks. Buy unsweetened cereals; consumers pay an excessive amount for pre-sweetened varieties. Try substituting honey or molasses in recipes calling for sugar (but use only ½ to ⅓ as much).

When eating something sweet, try to make sure it's nutritious. Try the following recipes for nutritious sweets:

Date Bars

8 oz. pitted dates
½ cup butter
1½ cup whole wheat flour
1¼ teaspoon baking powder
¼ cup honey
½ teaspoon salt
1 egg
¼ cup orange juice
1 tablespoon wheat germ

Beat egg into softened butter; mix all ingredients. Stir in chopped dates. Press batter into greased and floured 8″ or 9″ square pan. Bake at 375° F. about 25–30 minutes, until lightly browned. Cut into squares and remove from pan while still warm. Store in tightly covered container. Yield: 16–25 bars.

Pecan Balls

½ cup butter
½ cup solid vegetable shortening
1¼ teaspoons vanilla
2⅓ cups whole wheat flour
2 tablespoons milk
1½ cup finely chopped pecans
½ cup dried currants

Beat butter and shortening and vanilla. Mix in flour. Add currants and ½ of the pecans. Refrigerate for 2 hours. With floured hands shape dough into ¾″ balls. Coat balls in remaining pecans. Place on ungreased cookie sheet. Bake at 400° F. for 10 minutes. Cool on cookie sheet. Let stand 6 hours before serving. Store in tightly covered container.

Honey Cake

1½ cup whole wheat flour
½ teaspoon baking soda
½ teaspoon salt
3 tablespoons soft butter
¾ cup honey
1 egg
½ cup milk
¾ cup chopped almonds

Beat butter, honey and egg. Add flour, soda, salt. Then mix with milk until dry ingredients are moistened. Grease and flour loaf pan and sprinkle bottom of pan with some nuts.

Fold in remaining nuts. Bake at 325° F. for 45–50 minutes or until cake pulls away from sides of pan.

Additional Sources

Old-Fashioned Homemade Ice Cream Book, by Joyce and Christopher Dueker, Bobbs-Merrill, Indianapolis, IN, 1974.

The Natural Foods Sweet-Tooth Cookbook, by Eunice Farmilant, Doubleday & Co., Garden City, NY, 1973.

Natural Sweets and Treats, by Ruth Laughlin, Bookcraft, Salt Lake City, UT, 1973.

Cook With Honey, by Beverly Kees, Stephen Greene Press, Brattleboro, VT, 1973.

The Natural Sweet Tooth Breakfast, Dessert and Candy Cookbook, by Billie Hobart, Straight Arrow Books, San Francisco, CA, 1974.

#29 EAT WILD FOODS

One way to discover new eating experiences is to sample he vast assortment of wild foods available for the picking. dible fruits, nuts, roots, and greens can be found in every ural and suburban neighborhood. Wild foods can be found round vacant lots, deserted farms, pastures, stream beds and arshes, along roads and fences, and even in the back yard.

As a safety measure, do not eat unidentified wild plants. arts of certain plants are edible, while other parts of the same lant are poisonous. Most edible plants can be easily identified with a pocket guidebook (see References and Additional Sources). A few common-sense precautions are:

- Avoid plants with milky or colored juice. This includes milkweed, poison ivy, spurge, and poppy. There are exceptions however. For example, the young shoots of milkweed are edible, and even lettuce has milky juice.
- Avoid unknown white or red fruits. (Strawberries, apples, and tomatoes, although red, are safe and easily recognizable.)
- Avoid wild seeds; they are plant storage areas where toxins accumulate.
- Avoid mushrooms; they require expertise to identify those edible.
- Avoid any plant that tastes bitter, unless absolutely certain it is safe to eat.

Harvesting and eating wild plants can be pleasurable as well as nutritious. Growing in the purity of natural areas, wild plants are free of pesticides and are often more nutritious than store-bought vegetables. For example, the dandelion is rich in calcium, phosphorus, and vitamins B, C, and A. One ounce of dandelion contains 7,000 international units of vitamin A,[1] 2,000 to 3,000 more than the Recommended Daily Allowance. Also, wild plants are free for the taking and provide an opportunity to enjoy and learn from the natural environment.

One need not be an expert to take advantage of this free food. The wild edibles listed are found across the country and serve as a starting point from which one can learn more about nature's delicacies.

Wild Greens

Dandelions–This familiar vegetable is one of the best of the wild greens. The leaf edges are irregularly lobed and toothed. The yellow flowers mature into puffy white ovals, and the hollow stem contains a bitter milky juice.

In early spring, the young leaves are edible raw and can be used in salads. As the plants start to blossom, the leaves turn bitter. To prepare: Boil leaves and roots in salted water, changing it once. Serve with butter, margarine, or add to scrambled eggs and omelets.

Lamb's Quarters–This wild spinach is plentiful everywhere.

The entire young plant is edible and the tender tops are good into late summer. The pale green leaves have mealy looking undersides; the long, slender stalk can grow up to seven feet tall. When preparing lamb's quarters, add a vinegar sauce to enhance the flavor and preserve the vitamins. To prepare: Shred enough leaves to make 4 cups. Dice one onion and simmer in oil. Add ¼ cup vinegar; add salt and pepper to taste.

Wild Roots and Tubers

Cattail—These tall plants with brown, sausage-like heads grow in clusters throughout America's wetlands. Cattail roots and flower spikes are edible raw or boiled. In the fall, the roots can be washed, peeled, dried, and ground into meal. In the spring, the pollen-laden spike can be used to make cattail pancakes. Pancakes for 6 people can be made by sifting together 1½ cups of pollen, 1½ cups of whole wheat flour, 3 teaspoons of baking powder, 3 tablespoons of molasses, and ¾ teaspoon of salt. Beat three eggs and stir into 2 cups of milk; add 3 tablespoons of melted butter or margarine. After mixing with the flour combination, pour onto a hot, lightly greased frying pan. Serve with honey or molasses.[2]

Jerusalem Artichoke—These distinctively flavored tubers of the wild sunflower were cultivated by the Indians and are still raised commercially today. The egg-shaped leaves have hairy tops and sharp pointed tips. The numerous yellow flowers (2–3 inches in diameter) mature on slender stems. The plant is commonly five to ten feet tall.

The tubers may be harvested in late fall or winter if the ground is not frozen. Do not cook them at high temperatures because they toughen. One preparation method is to peel and dice a pound of tubers. Add the pieces to ¼ cup hot milk and stir, keeping them moist. Cover and cook just below boiling for about 10 minutes. Mix in 1 teaspoon of salt and 2 tablespoons of ground parsley. Sprinkle with paprika and serve.[3]

Wild Nuts

Black Walnut—The black walnut tree ranges from fifty to one hundred feet in height with a trunk two to six feet in diameter. The dark brown bark is furrowed and the oval-shaped leaves are from two to four inches long. In October the nuts ripen and fall to the ground. The nut is composed of an outer protective husk, a sculptured, bony shell, and a four-celled meat cavity. To remove the husks spread the nuts in the sun until they are partially dried. Place in a dry space to further dry for a month. Once dried, the nuts can be used in baking or eaten raw.

Beech Nut—The beech tree is easy to identify because of its smooth, blue-gray bark. The trees grow up to a hundred feet. The oblong leaves have pointed tips and wedged-shaped bases. Coarsely toothed, the leaves are thin and smooth. The nuts mature in October and can be easily opened by hand. The dried nuts can be used in baking and salads, or can be eaten raw.

Wild Fruits

Beside the common fruits like strawberries, raspberries, blueberries, blackberries, cranberries, grapes, cherries, plums, and crab apples, there are numerous less familiar fruits like the staghorn and sumac berries.[4] These make a drink much like lemonade when steeped in water and sweetened.

Mulberry—Mulberries ripen in the early summer and make delicious jellies and pies. The red mulberry is a small tree, generally 20 to 30 feet high with a trunk diameter from 1 to 1½ feet. It thrives in moist bottomlands and foothill forests, but can be found almost anywhere.

The tree's branches spread into dense, round domes. The heart-shaped leaves are toothed and irregular, with the bases ending in pointed tips. The ripe fruit resembles the blackberry in shape and color, becoming dark purple when ripe. To make mulberry jelly: Mash 8 cups of cleaned fruit in a large pot. Add ½ cup of water and slowly bring to a simmer. Boil rapidly for 10 minutes, then pour the mixture into a jelly bag to drip. Add 1 teaspoon of lemon juice to each cup of liquid. Stir in 4 cups of sugar and add 2 packages of pectin, boiling the juice until it falls in a sheet from the side of a spoon. Skim and pour immediately into hot sterilized glasses. Fill the last ½ inch of the jar with melted paraffin to seal; store in a dry, dark, cool place.[5]

Notes

1. "All the Weeds You Can Eat," by Robert Rodale, the Washington *Post,* December 5, 1974, p. E–18.
2. *Free for the Eating,* by Bradford Angier, Stackpole Books, Harrisburg, PA, 1966, p. 152.
3. Ibid., p. 138.
4. Ibid.
5. Ibid., p. 58.

Other References

"Concern Calendar 1975," Concern Inc., 2233 Wisconsin Ave., Washington, DC 20007.

Additional Sources

The Edible Wild: A Complete Cookbook and Guide to Edible Wild Plants in Canada and Eastern North America, by Berndt Berglund and Clare Bolsby, Charles Scribner's Sons, New York, NY, 1971.

Edible Wild Plants of the Western United States, by Donald Kirk, Naturegraph Publishing, Healdsburg, CA, 1971.

Stalking the Good Life, by Euell Gibbons, David McKay Co., New York, NY, 1971.

Feasting Free on Wild Edibles, by Bradford Angier, Stackpole Books, Harrisburg, PA, 1972.

Edible Wild Plants of Eastern North America, by Merritt Lyndon Fernold, Reed C. Rollins, Alfred Charles Kinsey, Harper & Row Publishers, Scranton, PA, 1958.

Indian Uses of Native Plants, by Edith Van Allen Murphey, Yankee Press, Inc., Dublin, NH.

Common Weeds of the United States, by Agricultural Research Service of the USDA, Peter Smith Publisher, Inc., Gloucester, MA, 1970.

One Thousand American Fungi, by Charles McIlvaine and Robert MacAdam, Peter Smith Publisher, Inc., Gloucester, MA, 1946.

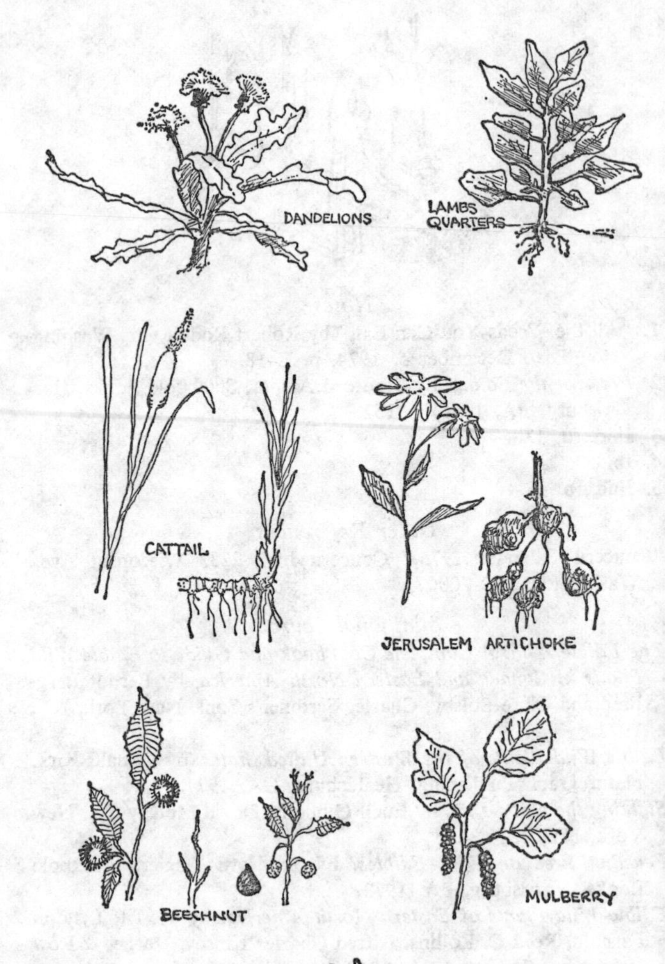
DANDELIONS
LAMBS QUARTERS
CATTAIL
JERUSALEM ARTICHOKE
BEECHNUT
MULBERRY

BLACK WALNUT

#30 LEARN TO PRESERVE FOOD

Food is wasted in fields, orchards, homes, and restaurants. Waste occurs while harvesting, transporting, storing, and retailing. Ripe fruit is left on trees, and vegetables decay on the vines. During growth, storage, and transportation, insects and rodents consume a large quantity of food.

Food is often wasted by processers who discard nutritious parts of plants, and by grocers who wait too long to reduce prices, allowing the produce to deteriorate to an inedible condition. Almost 15 per cent of supermarket bread is discarded by the stores. Restaurants and homemakers often serve large portions, much of which is not eaten and is thrown away. (In a 1973 survey in Tucson, Arizona, almost 10 per cent of the food purchased by sample households was discarded.) For economy and maximum utilization of a valuable commodity, food-storing practices should be improved.

Obtain the maximum yield from home-grown foods by harvesting at the proper time.[1]

- *Tomatoes:* Pick when red for the best flavor and nutritional value. Commercial markets use the mature-green stage; the next stage is pink, then red.
- *Onions:* Grow best in cool, moist, spring weather and mature in the warm, dry summer. Harvest when the tops turn yellow. Let dry in the sun for several hours. Cut off the tops and remove all dirt. Store in open mesh bags in a well-ventilated, dry place.
- *Beets:* Harvest when small to medium; large ones may be tough. Beets grow best in cool weather. Beet greens are best from young plants.
- *Lima beans:* Pick when the pods are green and the ends spongy.
- *Snap beans:* Pick when pods snap readily.
- *Broccoli:* Pick when the flowers show color.
- *Carrots:* For best eating, harvest when small, firm, and well-colored.

- *Cucumbers:* Pick before yellowing. A moderate size is best.
- *Peppers:* Harvest when mature, firm with good green color.
- *Summer squash:* Pick when approximately 6 inches long.
- *Swiss chard:* Harvest while crisp, tender, and not yet yellow.
- *Cantaloupe:* Are ripe when they have a fragrant smell. Softness at the blossom end does not necessarily indicate ripeness.
- *Sweet corn:* Harvest just as the silk blackens. Cook immediately; half of the sugar turns to starch within 24 hours after harvest.

Americans are almost exclusively dependent on energy-intensive methods of food storage. Creative, practical, healthful, low-cost techniques of food preservation can lead to a more self-sufficient household.

DRYING

Sun drying is especially recommended for fruits and can also be used for many vegetables. Those with no prior experience should start with the more commonly dried fruits: apples, apricots, prunes, and figs. Place the chopped fruits in a flat pan and cover with netting to protect them from insects. Fruit should be dried only on hot days so the drying process will be sufficiently fast. Bring the fruit indoors at night and store in a dry place until morning; otherwise, it will reabsorb moisture from the air. Drying times depend on type of fruit, temperature, relative humidity, and amount of ventilation. Properly dried fruit is leathery and will not yield a single drop of moisture when cut in half.

Herbs can also be easily dried. Cut plants when leaves are still green and tender. Spread herbs on a large piece of wrapping paper and cover with another piece of paper to keep out light. The herbs should be placed in an attic that is warm but below 100° F. and left for 4 or 5 days to dry.

SMOKING

Smoking is a variation of drying; both rely on the removal of moisture to discourage growth of the micro-organisms that lead to spoilage. While the process has been utilized for centuries for curing meat, smoking can also be used to preserve fish, cheese, and certain vegetables. Since commercially smoked products are usually treated with chemicals (e.g. sodium nitrite), home smoking is advantageous from a health standpoint.

A simple home smoker can be made from an old refrigerator. Place a pot or tub large enough to serve as a fire box on the floor of the refrigerator. Food to be smoked is placed on racks in the refrigeration cavity. Holes must be drilled to allow smoke to pass through the food and out of the smoker. By selectively covering the holes, the temperature and the flow of smoke through the smoking chamber can be controlled. A smoker can also be improvised from an old 55-gallon oil drum.

Burning hardwoods lend the best flavor to smoked foods. Since smoking alone is usually not sufficient to prevent spoilage, foods must be treated with salt, sugar, or some other spice prior to smoking. For specific smoking recipes, see *Home Preserving Made Easy* by Vera Gewanter and Dorothy Parker.[2]

ROOT CELLAR STORING

Most fruits and vegetables can be effectively stored in a root cellar. A root cellar is a cool, dry place with temperatures between 30° and 50° F. Air should be allowed to circulate and must not be too dry.

A section of a basement can be converted into a root cellar. It should be partitioned off from the rest of the basement, have one or preferably two windows, and be insulated and equipped with shelves. Fruits and vegetables should be stored separately since odors from certain vegetables, such as onions and turnips, adversely affect the flavor of fruits. Part of a tool shed could also be converted into a root cellar. A house without a basement might have space outside to build a hatch jutting out from the house. The walls of the hatch should be partly underground, the floor should be made of packed soil, and the door should be made accessible by a few steps.

As a last resort, partially bury an old trunk, barrel, or plastic garbage can. The part of the root cellar above ground should be covered with straw and dirt to help protect against temperature extremes and moisture. Bury the receptacle in such a manner that the opening is easily accessible. Apartment dwellers might make a food cage or an outside cupboard on a fire escape or balcony. In cold climates, leaves and sand hold off frost and rot while at the same time keep dry air from evaporating juices from fruits and vegetables. Food will not spoil unless it freezes and then thaws. Although warmer temperatures will shorten storage time, they will not damage fruits and vegetables.[3]

PICKLING AND SALTING

Fish, fruits, vegetables, nuts, and cheeses can be pickled. Pickling is a simple process that does not require steam pressure, very high heat, or any kind of refrigeration. It agreeably alters taste, consistency, and sometimes the color of food, and safeguards against mold and bacteria.

To Salt Fish[4]

1. Clean and wash fresh fish thoroughly. Scrape off scales and remove slime by washing in water with 1 teaspoon vinegar. Remove the head, fins, and tail.
2. Spread a layer of salt over the bottom of a large wooden container (use coarse or rock salt).
3. Place a layer of fish, flesh side up, in the container.
4. Completely cover the fish with salt.
5. Repeat steps 3 and 4 until the container is almost full. Keep the layers even and compact. Put the last layer of fish skin side up.
6. Top with a layer of salt.
7. Put a weight on the fish to keep them under the salt. Cover tightly. At least ten to fourteen days are required.

8. The fish may be left for a longer time if stored in a cool place.
9. To use, soak overnight in clean, cold water.

FREEZING[5]

A Guide to Freezing Fruits and Vegetables

Vegetables

Select young, tender, garden-fresh vegetables.

Wash vegetables in cold water, trim according to directions below.

Blanch. (Parboil or pour boiling water over food, then drain and rinse with cold water in order to stop enzyme action, which is aging the vegetables.) This applies to all vegetables with exceptions that will be noted. Each pound of prepared vegetables requires at least one gallon of boiling water. According to microwave manufacturers, foods can be blanched in the ovens.

Use blancher or large kettle, with cover, into which wire basket fits.

Count blanching time when water returns to boil after adding vegetables.

Cool immediately by plunging vegetables into ice water.

Dry pack in plastic containers or bags, allowing head space.

Cook frozen.

Vegetable	*How to Prepare for Freezing*
Beans, green or wax	Trim ends. Cut into 1-inch or 2-inch pieces or french. Blanch 1½ minutes.
Beans, lima	Remove from pods. Sort for size. Blanch 2 to 3 minutes, depending on size.
Beets	Small whole: Peel and blanch 5 minutes. Mature large: Cook until tender, 45 to 50 minutes. Peel, slice or dice.
Broccoli	Trim off woodiness and large leaves. Cut into serving pieces. Split large stalks to ½-inch thickness. Blanch 2 to 3 minutes.

Brussels sprouts	Trim outer leaves. Blanch 3 to 4 minutes.
Cabbage	Suitable only for cooked dishes. Trim coarse outer leaves. Coarse shred or separate leaves. Blanch 1½ minutes.
Carrots	Scrape, cut in ⅓-inch dice. Blanch 2½ minutes. Small whole carrots: Blanch 3 to 5 minutes.
Cauliflower	Break into serving pieces. Blanch 3 minutes.
Celery	Trim, cut into 1-inch lengths. Blanch 3 minutes. Suitable only for cooked dishes.
Celery root	Wash and trim. Cook until almost tender. Peel and slice.
Chestnuts	Boil, drain and shell.
Chives	Chop. Do not blanch.
Corn	Husk and de-silk. On the cob: Blanch 5 to 8 minutes. Cut corn: Blanch on cob 1½ minutes. Chill, then cut kernels from cob.
Eggplant	Peel, cut in ⅓-inch slices or dice. Blanch 4 minutes, then dip in ½ cup lemon juice mixed with 2½ pints cold water.
Fennel	Trim, cut in 1-inch lengths. Blanch 3 minutes.
Green peppers	Wash, remove seeds and stem; chop. Do not blanch.
Kohlrabi	Small roots, 2 to 3 inches in diameter. Cut off tops, peel and dice. Blanch 1½ minutes.
Okra	Cut off stems. Small pods: Blanch 2 minutes. Large pods: Blanch 3 minutes.
Onions	Peel and chop. Do not blanch.
Parsley	Chop. Do not blanch.
Parsnips	Cut off tops, peel. Cut into ½-

	inch cubes or slices. Blanch 2 minutes.
Peas	Black-eyed: Shell, blanch 2 minutes. Green: Shell, blanch ½ to 1 minute.
Pumpkin	Seed and quarter. Cook until soft. Remove pulp and mash.
Rutabagas	Cut off tops, peel and dice. Blanch 1 minute.
Spinach	Blanch 1½ minutes.
(beet greens, chard, collards, kale, mustard greens, turnip greens)	Collards: 3 minutes. Other greens: 2 minutes.
Squash, summer varieties	Cut in ½-inch slices. Blanch 3 minutes.
Squash, winter	Cut into pieces, seed. Cook until tender. Remove pulp and mash.
Sweet potatoes	Cook until almost tender. Cool and peel. Dip whole or sliced in solution of ½ cup lemon juice to 1 quart water for 5 seconds or add 2 tablespoons orange or lemon juice to each quart of mashed.
Tomatoes, puréed	Trim and quarter. Simmer 5 to 10 minutes. Press through sieve. May be seasoned with 1 teaspoon salt to each quart juice.
Tomatoes, stewed	Trim and quarter. Simmer until they can be forced through a strainer. May be seasoned with salt, 1 teaspoon per quart.
Turnips	Peel and quarter. Cook 10 to 20 minutes. Peel, cut into ½-inch cubes. Blanch 2 minutes.

Fruits

Select firm texture, mature fruits. Prepare to freeze immediately after harvesting. Certain varieties freeze more successfully than others, but if freshly picked and carefully handled, most fruits will freeze satisfactorily.

Wash thoroughly in cold water. Cut and trim according to directions below.

Pack in "Cold Syrup," "Sugar Pack" or "Unsweetened Pack."

Cold Syrups

Per Cent	*Sugar in Cups*	*Water in Cups*	*Yield*
30	2	4	5
40	3	4	5½
50	4¾	4	6½

Cold Syrup

Follow the above table for per cent of cold syrup called for in directions, dissolving sugar in cold or hot water. If hot water is used, cool syrup. The syrup may be made up beforehand and refrigerated. To insure that the syrup covers all the fruit, press a crumpled piece of freezer paper on top and press down fruit into syrup before sealing lid.

Sugar Pack

See instructions below for amount of sugar needed for each fruit. Mix sugar gently with fruit until it is dissolved. Press crumpled piece of freezer paper on top of fruit to keep fruit covered. Make a note on the container of amount of sugar used.

Dietetic Pack

Sugar substitutes can be used for those on special diets. Write to the manufacturer of the artificial sweetener for specific directions, if they do not appear on the package.

Unsweetened Pack

Dry pack or cover with water containing required amount of ascorbic acid, given in directions.

Ascorbic acid is needed to prevent some light fruits—those shown below—from darkening and losing flavor. However, if cold fresh fruit is cut directly into cold syrup and covered with it, then frozen at once, there should be little darkening without the ascorbic acid.

For dry pack use rigid containers or plastic bags. "Head" room is not needed. Fruits frozen by other methods should be packed in rigid containers, allowing a half inch to one and a half inches of headroom.

Fruits to be served without cooking should not be completely thawed. A few ice crystals in them keep them from going limp and enhance their flavor.

Fruits	*How to Prepare for Freezing*
Apples (with the exception of Red Delicious)	For pies: Peel, core, slice. Drop in boiling 30 per cent syrup, scald 2 minutes. Lift from syrup, drain and pack.
Apricots	Plunge into boiling water for 30 seconds to loosen skins; peel. Cut in half or slice into 40 per cent cold syrup with ¾ teaspoon ascorbic acid per quart syrup. Purée: Quarter, heat to boiling in just enough water to prevent scorching; sieve. 1 cup sugar for each quart and ¼ teaspoon ascorbic acid dissolved in ¼ cup water.
Berries (blackberries, blueberries, raspberries, boysenberries, cranberries, elderberries, gooseberries, strawberries, huckleberries, dewberries, loganberries, youngberries and currants)	All may be frozen dry pack method. Sort and stem. Spread on trays until frozen, then pack. Syrup pack: Use 50 per cent cold syrup for gooseberries, cranberries and currants. Use 40 per cent syrup for all others.
Cherries, sour	Sort, pit if desired. Mix with dry sugar (1 pound sugar to 5 pounds cherries) until dissolved.
Cherries, sweet	Sort, pit if desired. Cover with 40 per cent cold syrup mixed with ½ teaspoon ascorbic acid per quart of syrup.
Figs	Sort, stem, peel and slice. Use 30 per cent cold syrup or dry pack. Dry pack dried figs.

Grapes	Stem. Leave seedless grapes whole. Halve and seed others. Pack in 30 per cent cold syrup.
Melons	Seed; ball, cube or slice into 30 per cent cold syrup. Add 1 teaspoon lemon juice to each cup syrup.
Nectarines	Peel, if desired. Sort, pit; quarter, halve or slice into 40 per cent cold syrup with ½ teaspoon ascorbic acid per quart syrup.
Peaches	Sort, pit and chill. Peel without scalding, if possible. Halve, quarter or slice into 40 per cent cold syrup with ½ teaspoon ascorbic acid per quart syrup.
Pears	Peel, halve or quarter and core. Heat in boiling 40 per cent syrup 1 to 2 minutes depending on size of fruit. Drain and cool. Pack in 40 per cent cold syrup with ¾ teaspoon ascorbic acid per quart syrup.
Plums	Sort. Halve or quarter if desired. Pit. Pack in 40 per cent cold syrup with 1½ teaspoon ascorbic acid per quart syrup. Italian plums must be dry packed.
Rhubarb	Trim, cut into desired lengths. To retain flavor heat in boiling water 1 minute and quick cool. Pack in 40 per cent cold syrup or dry pack.

(Reprinted with permission from the Washington *Post*)

CANNING

Canning is a practical method of preserving surplus food. The cost of the canning operation varies tremendously, depending on whether food is home-grown, and on the cost of cooking energy and canning materials. Group canning is generally far less expensive than individual home canning. Cost for canning a quart of tomatoes at home ranged from 4.3 cents if jars were on hand and the tomatoes were acquired free of cost to almost 51 cents if jars and tomatoes were bought.[6] There are only small savings if jars and produce have to be purchased. Savings are further reduced if commercially canned foods can be bought in case lots at discount prices.

Some considerations for home canners are:

1. Adequate storage space must be available.
2. Canning without exact instructions may result in food waste and illness due to food spoilage.
3. Canning requires high standards of hygiene. All equipment must be carefully washed in hot soapy water and rinsed thoroughly.
4. Jars must be made of tempered glass and there must be no chips or defects round the rims.
5. Boil home canned foods 10 minutes or more if not absolutely certain of sanitation of canning methods.
6. It is economical to can only the amount that can be used in a reasonably short length of time.
7. It is not necessary to add syrup when canning most fruits; the natural sweetness of fruits is usually sufficient.

Shared canning is not only a pleasant social experience, but is also much more economical in terms of effort and energy. Group canning goes twice as fast and can bring family and community together in a creative, useful activity.

Notes

1. "The Care and Feeding of Home-Grown Vegetables," by Tom Stevenson, the Washington *Post,* July 13, 1975, p. F-7.
2. *Home Preserving Made Easy,* by Vera Gewanter and Dorothy Parker, Viking Press, New York, NY, 1975.
3. Ibid.
4. "To Salt Fish," A University of Alaska Pamphlet, from *Public Works,* edited by Walter Szykita, Links Books, New York, NY, 1974.
5. "After Growing, Freeze It," the Washington *Post,* July 31, 1975, p. F-2.
6. *Canning and Freezing–What Is the Payoff?,* talk by Evelyn H. Johnson at the 1976 National Agricultural Outlook Conference, Washington, DC, November 20, 1975.

Additional Sources

Ball Blue Book, available from Ball Brothers Co., Inc., Muncie, IN 47302.

Kerr Home Canning and Freezing Book, available from Kerr Glass Mfg. Corp., Consumer Products Division: Box 97, Sand Springs, OK 74063.

The Pleasure of Preserving and Pickling, by Jeanna Lesem, Alfred A. Knopf, New York, NY, 1975.

Stocking Up: How to Preserve the Foods You Grow Naturally, by the editors of Organic Gardening and Farming, Rodale Press, Emmaus, PA, 1973.

Complete Guide to Home Canning, Preserving and Freezing, by the U. S. Department of Agriculture, Superintendent of Documents, U.S.G.P.O., Washington, DC 20402.

Dry It–You'll Like It, by Glen MacMamiman, Montana Books, Seattle, WA, 1974.

Complete Book of Home Storage of Vegetables and Fruits, by Evelyn Loveday, Garden Way Publishing Co., Charlotte, VT, 1972.

How to Make Jellies, Jams and Preserves at Home, a Department of Agriculture booklet, Superintendent of Documents, U.S.G.P.O., Washington, DC 20402.

Putting Food By, by Ruth Hertzberg, et al., Steven Green Press, Brattleboro, VT, 1975.

Freezing and Canning Cookbook, by the food editors of Farm Journal, Doubleday & Co., Garden City, NY, 1973.

Home Canning of Fruits and Vegetables, Home and Garden Bulletin #8, Superintendent of Documents, U.S.G.P.O., Washington, DC 20402, 1969.

Storing Vegetables and Fruits, Home and Garden Bulletin #119, Superintendent of Documents, U.S.G.P.O., Washington, DC 20402.

#31 CONSERVE NUTRITIONAL VALUE OF FOOD

Quality produce with high nutritional value begins on organic farms where chemical fertilizers and pesticides are not used and fruit is allowed to ripen on trees and vines. The mineral and vitamin contents are maximum when fruits are ripe. For easier handling during transport, corporate farmers routinely harvest before produce is completely ripe. This may aid efficiency, but it produces low nutritive values. The ideal situation is vine- and tree-ripened produce with rapid transportation to the consumer. This could be attained if small farms were close to residential areas or if home owners farmed their yards and city dwellers cultivated their parks and vacant lots.

Further deterioration in valuable food elements occurs at home. Improper storage, preparation, and cooking robs foods of vitamins and minerals. For the best food value, buy wisely, cook properly, and be conscious of the nutritional benefits of various foods.

Shop in the morning when vegetables are fresh. Choose those that have not been trimmed. Store fruits and vegetables according to the requirements of each for temperature, humidity, and ventilation.

STORING FOOD[1]

- Wash and dry leafy vegetables and store at near-freezing temperature and high humidity. At room temperature and in bright light, up to 50 per cent of the folic acid, vitamin B_2, and vitamin C can be lost in one hour.
- Keep green peas and lima beans in their pods until ready to use. If bought shelled, keep them in the refrigerator.

- Orange juice reconstituted, canned, or fresh will keep its high vitamin C level for several days if refrigerated. To get the maximum nutritional value, peel and eat fresh oranges.
- Berries are a good vitamin C source. They must be handled carefully and should not be capped until ready to be eaten. Bruised and decapped berries lose vitamin C quickly.
- Keep canned foods in a cool, dry place. Rotate with new supplies, since vitamin quality decreases with age. Canned meat should have only a three- to four-month shelf life.
- Frozen foods lose vitamin C by one-third to three-fourths over one year's time. Thawed and refrozen foods lose nutritive values; the flavor is affected and spoilage can result.

PREPARING VEGETABLES[2]

- Do not peel vegetables; the peel is high in nutrients and provides roughage. Clean by rinsing and rubbing. Trim bruised areas with a sharp blade. Remove damaged leaves, bruised skin spots, and inedible parts such as stems and ribs from collard, kale, and turnip greens. Use cabbage cores and broccoli leaves; both are high in vitamins.
- Whirl salad greens in a large container, add salad oil, and toss so that each piece glistens. These measures prevent the loss of soluble vitamins and minerals.
- Save all scraps and peels from vegetables. Boil them in a cloth sack and save the liquid. Save the water from soaking dried beans, peas, and lentils.
- Do not soak vegetables, add soda, or store in a wet condition. Do not leave in open air unnecessarily. These measures preserve vitamins.
- Chill fruits and vegetables before cutting.

COOKING VEGETABLES[3]

- Eat raw fruits and vegetables as often as possible. Cooking can destroy vitamins.
- DO NOT OVERCOOK–nutrients are lost.
- Pressure cooking, steaming, and broiling are preferable to boiling. Rapid initial heat and a shorter cooking time preserves nutrients.
- When baking or frying, cover the pan with a lid. Contact with air destroys vitamins.
- Boiling is the most destructive cooking method. Seventy per cent to 100 per cent of natural sugars, minerals, and vitamins can be lost. Aromatic oils are dissolved, which results in a loss of flavor. When boiling use a minimal amount of water and use the remaining water in soups.
- Cook frozen vegetables without thawing in a minimal amount of water.
- Add salt after the food is cooked. Salt increases nutrient loss in water.

ENHANCING NUTRIENT VALUE OF FOOD

- Use dried skim milk in soups, milk drinks, and bread. This adds protein, vitamin B_2, and calcium without increasing the fat content.
- Add brewer's yeast, wheat germ, soy flour, peanut, and cottonseed flours to wheat flour for bread, and increase the protein by five to ten times.
- Substitute blackstrap molasses for sugar in formulas, bread, and in desserts. It has ten times the nutritive value of plain molasses. One tablespoon has as much calcium as a glass of milk and as much iron as nine eggs.

Notes

1. *Conserving the Nutritive Values in Foods,* USDA Booklet, Superintendent of Documents, U.S.G.P.O., Washington, DC 20402.

2. *Let's Cook It Right,* by Adelle Davis, Harcourt, Brace & Jovanovich, New York, NY, 1970, pp. 308–17.
3. Ibid.

Additional Sources

Refer to entries #25 and #37.
The Greengrocer Cookbook, by Joe Carcione, Celestial Arts, Millbrae, CA, 1975.

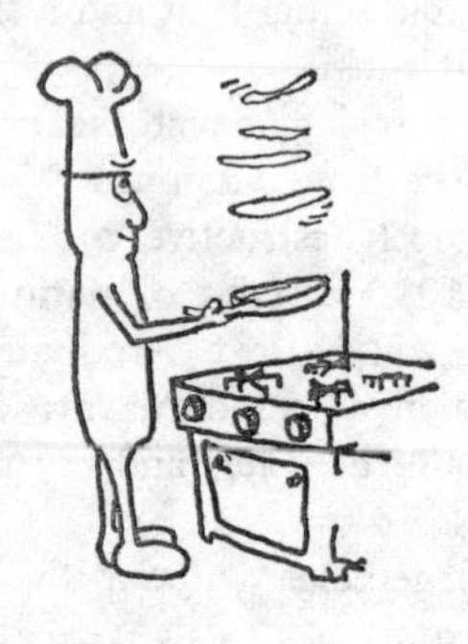

#32 BE AWARE OF AGRIBUSINESS

Consumers should be aware of the flaws in American food marketing. High food costs would be more acceptable if farm workers and small farm owners were benefiting, but this is not the case. In the past twenty-five years, the number of cultivated acres in America has remained constant, but the number of people living on and working the soil has dropped by 3 million, or 50 per cent.

Since World War II, the independent family farm and small grocery store has been replaced by the corporate conglomerate and the chain supermarket. It may surprise readers to know that Greyhound produces turkey, ITT bakes bread, and Tenneco markets vegetables. Even former Secretary of Agriculture Earl Butz has admitted that farming is no longer a way of life—it's a business. Corporations are expanding horizontally by incorporating similar processes, and vertically by entering agriculture and food marketing.

Consumers may wonder why giganticism is bad, especially since food production has increased through the years. It would seem that anything that achieves efficiency and increases food supplies would benefit the U.S. and world populations. However, monopolies lead to reduced competition, higher advertising costs, excess profits, price fixing, and curbing of innovative ideas.

One adverse effect is that in order to sell their produce, farmers must comply with artificially imposed terms. Middlemen separate the farmer from the consumer, controlling sales and pricing procedures. It is not uncommon for one firm to control 60 to 70 per cent of the marketing of a particular produce.

The actual farming practices are being mechanized to an even greater extent—quite possibly to a state of diminishing returns. Energy (primarily petroleum and natural gas) in the form of fuel, fertilizer, pesticides, and herbicides is being used in large amounts to increase crop yields.

The use of fertilizer has increased ten times since 1946. Fossil fuel consumption to power farm equipment has grown 50 per cent. The 230 per cent increase in corn yield per acre from 1945 to 1970 resulted from a 310 per cent increase in fuel consumption. In fact, the production of one bushel of corn now requires one gallon of gasoline.

In the early 1900s, the American farmer burned less than one calorie of fuel energy to produce one calorie of food energy. Today, farmers use about ten calories to produce one calorie of food energy. It is not uncommon for many "backward" agrarian societies to obtain from five to fifty food calories for each calorie invested in food production.

Cattle and poultry are confined to pens and lots for intensive feeding before butchering. Growth-stimulating drugs, such as DES, and disease-fighting chemical additives lace the animals' diets. These additives affect the meat's taste and texture and can possibly cause cancer.

Consumers should not feel totally helpless in matters of food production. Everyone should be aware that manure, compost, and other organic fertilizer sources can supplement chemical fertilizer (see entry #44). Alternative energy sources such as solar, wind, and methane gas from animal waste, should be developed as replacements for fossil fuels (see entries #12–#14).

Individual actions can be taken, such as growing fruits and vegetables at home to ensure one's family of good produce while reducing grocery bills (see Section IV). With increasing interest in natural foods and organic gardening, a number of non-profit food stores and co-ops have been started. Many of these stores aim to reduce marketing costs by eliminating supermarket frills, such as pre-packaging and excessive advertising. Wholesome food, bulk quantities, and reduced prices characterize many of these stores (see entry #33).

Reference

All statistics taken from: *Food for People, Not for Profit,* edited by Catherine Lerza and Michael Jacobson, Ballantine Books, New York, NY, 1975.

Additional Sources

Eat Your Heart Out, by Jim Hightower, Crown Publishers, New York, NY, 1975.

Hard Tomatoes, Hard Times, by Jim Hightower, Schenkman Publishing Co., Cambridge, MA, 1973.

From the Ground Up: Building a Grass Roots Food Policy, from CSPI Publications, 1757 S St., NW, Washington, DC 20009.

The People's Land, by Peter Barnes, Rodale Press, Emmaus, PA, 1975.

National Farmers Union
1012 14th St., NW
Washington, DC 20005

or

NFU
Box 39251
Denver, CO 80239

National Catholic Rural Life Conference
3801 Grand Ave.
Des Moines, IA 50132

Center for Rural Affairs
PO Box 405
Walthill, NB 68067

#33 ORGANIZE A FOOD CO-OP

Operating a food co-op is conducive to community involvement and good nutrition. As a democratic society, people openly participate in establishing a store. There is no set number of members; only a few are paid, and most are volunteers.

A case study of the Bethesda Avenue Co-op illustrates how such an enterprise begins. Citizens with varying interests in food, economy, and health began meeting regularly. Rounding up neighbors and community acquaintances, this core of enthusiasts discussed competent management and decided operational details:

- Organizational structure (whether to be a co-operative, non-profit corporation, or workers' collective).
- Prepaid orders versus store-front inventories.
- Use of volunteer labor versus professional management.
- Nutritional orientation (natural foods versus conventional packaged goods).
- Goals (consumer savings and community organization).

Some workers began serving as volunteers in similar stores in other neighborhoods. They learned about accounting systems, pricing methods, and supply sources. For example, they discovered how a non-profit community warehouse bought produce from a large farmer's market and milled its own flour. They found a non-profit trucking collective that would deliver to the co-op.

Guidelines were set to ensure consumer and community control, as this store would be created by the people who worked for it. (Stores with non-community based workers frequently have a high turnover in their work force.) Anyone who supported the co-op was a member, and three hours of work per month entitled a member to vote. A board of directors was formed, evenly composed of salaried employees and volunteer workers.

Everyone worked to expand the mailing list. To obtain operating funds and personal involvement, pamphlets were sent to civic associations, religious groups, and individuals. They found a space for store rental near a shopping area, convenient to lower-, middle-, and upper-class neighborhoods.

To raise money, the organizers gave community dinners that featured nutritious, non-meat dishes. They collected additional revenue from raffling off homemade specialties and recipes. Certificates were sold, redeemable after six months for food from the co-op. A few citizens even gave short-term loans.

About $1,500 should be enough to stock a small store with the basics: grains, bread, nuts, beans, fresh produce, and cheese. A co-op store can be opened with around $5,000. The process of expanding stock and volume will be slow. It is necessary to keep raising funds to achieve the goal: a one-stop shopping place.

Low prices, peak quality produce, and excellent cheese will provide a start for pleasing initial customers, who then can spread the word and possibly lend financial support. Information on salaries, cost, and mark-up should be public information. The co-op should give confidence to the consumer while providing a sense of buyer control.

People who take pride in their community store will feel more positively toward the community and are likely to become more actively involved in its activities.

A few other ideas might be helpful if one is interested in alternative marketing procedures. Locate and support alternative markets in the community–farmer's markets, roadside stands, pick-your-own enterprises, and other co-ops. Collect information and published literature on existing operations to gain knowledge on getting started and to become familiar with problems that new co-ops face.

It might be helpful to discuss the co-op idea with a local consumer affairs office, community action agency, or local farmer organization. Finally, be sure to publicize the benefits of the co-op, especially lower prices and fresher produce.

Reference

An interview with a member of the Bethesda Avenue Co-op, Bethesda, MD, by Cheryl Tennille, a CSPI volunteer.

Additional Sources

Food Co-op Handbook: How To By-Pass Supermarkets and Control Price and Quality of Your Food, by Co-op Handbook Collective, Houghton Mifflin Co., Boston, MA, 1975.

Food Co-ops: An Alternative to Shopping in Supermarkets, by William Ronco, Beacon Press, Cambridge, MA, 1974.

The Cooperative League of the U.S.A.
Suite 1100
Washington, DC 20036
(materials available covering entire range of co-op information)

Community Services Administration
1200 19th St., NW, Rm. 332
Washington, DC 20506
(write for free reports on co-op operation and the National Consumer Directory)

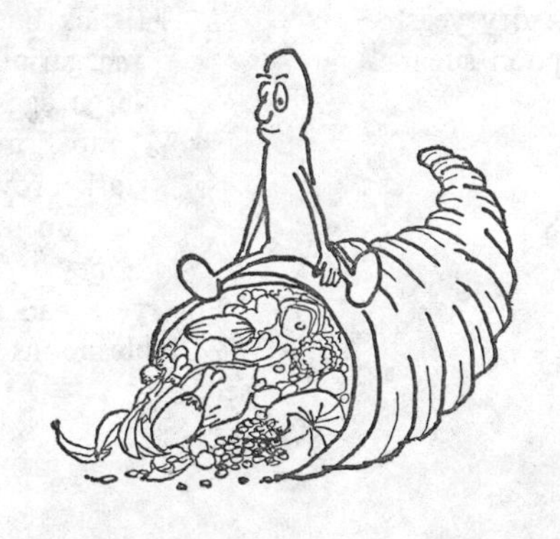

Self-Help Action Center
11013 South Indiana Ave.
Chicago, IL 60628

Food Advocates
2288 Fulton St.
Suite 200
Berkeley, CA 94704

#34 BAKE BREAD

Home-baked bread is both economical and satisfying. Made in batches of four to six loaves at a time, it is worth the effort and can be wrapped and frozen when thoroughly baked. A simple method follows for making four varieties of protein-rich bread. Steps A and B are the same for all four varieties, with the exception of rye bread which has caraway seeds added to step B. The different types are made by varying the flours and ingredients of C.

Recipes for 3 large loaves weighing close to
2 lbs. each, in pans 5¾″ × 9¾″ × 2¾″

A: 1 cup warm water
2 pkgs. dry yeast
1 teaspoon sugar

B*: 3 cups hot water
⅔ stick butter or margarine
3 large eggs
¼ cup firmly packed dark brown sugar or ¼ cup molasses or honey

* For rye bread only, include 1½ tablespoons caraway seeds.

C for White/Soy/Wheat Germ Bread:

9 cups unbleached white flour
3 cups soy flour
1 cup wheat germ
1½ cups instant dry milk
2½ teaspoons salt
White flour for kneading

C for Rye/Whole Wheat/White Bread:

6 cups rye flour
4 cups whole wheat flour
2 cups unbleached white flour
1½ cups instant dry milk
2½ teaspoons salt
Whole wheat flour and/or unbleached white flour for kneading

C for Whole Wheat/Oatmeal/White Bread:

7 cups whole wheat flour
3 cups quick rolled oats
3 cups unbleached white flour
1½ cups instant dry milk
2½ teaspoons salt
Whole wheat and/or white flour for kneading

C for Whole Wheat/Soy Bread:

9 cups whole wheat flour
3 cups soy flour
1½ cups instant dry milk
2½ teaspoons salt
Whole wheat flour and/or unbleached white flour for kneading

Directions for all four varieties:

Let step A ingredients get foamy in a bowl about 1-pt. size.

Mix step B ingredients well, in a larger bowl, about 1½-qt. size, adding eggs last.

Stir step C ingredients well in bowl, at least 4-qt. size.

Combine A and B, then add to C and beat by hand to mix well. Cover bowl with waxed paper and put in warm place to rest for about 10–15 minutes.

Turn onto floured board and knead 10 minutes, adding flour as necessary in order to knead dough.

Wash, dry, and thoroughly grease bowl, place dough in it and turn dough over to grease top. Return to warm place to double in bulk. A good test is to poke one's finger in the dough and if the dough does not fill the hole left by the poking finger, it is ready for the next step.

Punch the dough down thoroughly with fist, revolving the bowl while punching down, to pull dough from side into center.

Let dough rest 10 minutes.

Turn onto floured board. Cut into three portions. Shape into loaves by patting and flattening by hand into a rectangular shape and then rolling into loaf. Place into well-greased loaf pans, cover with waxed paper, and put in a warm place to rise till double. Heat oven to 360° F. When dough has risen, it can be moistened gently and seeds pressed gently into top, if desired.

Bake for 40 minutes or until well browned.

Cool thoroughly on racks; when cold, wrap and store in refrigerator or freezer.

Reference

Recipes from Dorothy Moursund, CSPI volunteer.

Additional Sources

"Why Bake Bread at Home?" by Jean Mayer, the Washington *Post,* May 17, 1975.

The Bread Book, by Carlson Wade, Drake Publishing, Inc., New York, NY, 1973.

Beard on Bread, by James Beard, Random House, New York, NY, 1973.

The Tassajara Bread Book, by Edward E. Brown, Shanbala, Berkeley, CA, 1970.

The Complete Sourdough Cookbook, by Don Holm, Caxton Printers, Ltd., Caldwell, ID, 1972.

Cooking with Whole Grains, by Mildred Ellen Orton, Farrar, Strauss and Giroux, New York, NY, 1971.

The Complete Book of High-Protein Baking (based on principles of *Diet for a Small Planet*), by Martha Ellen Katz, Ballantine Books, New York, NY, 1975.

Wheat for Man . . . Why and How (with recipes explicitly for use of stone-ground, whole wheat flour), by Vernice G. Rosenvall et al., Bookcraft, Salt Lake City, UT, 1966.

#35 PREPARE VARIOUS FOOD PRODUCTS

Other ways to cut food bills and channel personal creativity is to make products at home that one normally buys from the store, including: cereals, yogurt, sprouts, party dips, peanut butter, and salad dressings.

Breakfast cereals are an overpriced store item that could be made easily at home–and they would taste better too! A number of grains, nuts, and fruits can be combined to make nutritious foods to start the day or to eat as snacks. Try combinations using these ingredients, toasted or raw:

Rolled Oats	Almonds	Raisins
Wheat Germ	Sunflower Seeds	Apples
Wheat Flakes	Sesame Seeds	Dates
Coconut	Honey	Cinnamon
Peanuts	Dry Milk	Bran

Yogurt is a healthful and delicious dairy product with limitless versatility as a snack, salad dressing, dessert, or main-dish complement. It is simple to make at home using low fat or whole milk; a quart of homemade yogurt can cost less than eight ounces of any store-bought brand. Homemade incubators are not always efficient, so one might want to invest in an inexpensive incubator that regulates the temperature and humidity needed to grow the yogurt culture. Anyone who enjoys eating yogurt will receive a quick return on the investment.

Sprouts are excellent in salads or with a number of dishes.

In a quart jar, cover two tablespoons of alfalfa seeds with warm water and soak for eight hours. Drain the liquid and rinse the seeds. Place the jar on its side and rinse the seeds a couple of times each day. After a few days, the seeds will have sprouted and will be ready to eat. Keep them refrigerated.

Cottage cheese mixed with chopped onions and herbs will make an excellent party dip. Mix cottage cheese with raisins, rice, coconut, and cinnamon for a delicious dessert.

Make peanut butter by chopping or crushing peanuts to a fine consistency. Add oil or salt to suit personal taste and store in the refrigerator.

Any oil-and-vinegar salad dressing can be enlivened with additions of herbs, lemon juice, or hot mustard. Mixing mayonnaise with onions, tomato sauce, honey, and lemon juice makes a great French dressing.

There are limitless possibilities for using imagination to create products at home that are normally purchased at the store. Read the labels of products and take it from there–the recipes are often provided.

A few other recipes that you might want to try:

Mayonnaise

Mix 1 cup evaporated milk, 1 cup salad oil, ½ teaspoon salt, ¼ teaspoon paprika, blend for one minute. To the blended mixture gradually add ⅓ cup lemon juice, or enough to thicken it to desired consistency. Let stand till thick (about ½ hour) and till air bubbles appear. Blend again till smooth and refrigerate in covered quart containers.

Low-calorie Slaw Dressing

Combine ½ pint low-fat yogurt (lemon or lime yogurt can be used), 2 tablespoons lemon juice, and sweeten to taste. Makes enough for small head of lettuce.

French Dressing

Combine 1 cup oil, ⅓ cup honey, ¾ cup catsup, ⅓ cup chili sauce, ⅓ cup lemon juice, 1 teaspoon paprika, 1 teaspoon salt, ½ cup diced onion, 1 clove minced garlic. Put in quart jar and shake. Keep refrigerated. Yield: 3 cups.

Granola Cereal

- 5 cups oats
- 1 cup coconut
- ½ cup chopped cashews or slivered almonds
- ½ cup wheat germ
- 1 teaspoon ground cinnamon
- ½ cup honey
- ⅓ cup vegetable oil
- 1 teaspoon vanilla
- 1½ cups chopped dates or raisins
- ½ cup sunflower seeds
- 1 tablespoon sesame seeds (optional)

Preheat oven to 350°. Spread oats in a 13″ × 9″ pan and heat 10 minutes. Combine all dry ingredients except fruit in a large bowl. Stir in honey, oil and vanilla. Mix until dry ingredients are well coated. Spoon into same pan and bake 30 minutes more, stirring often to brown evenly. After cool, mix in raisins or dates and store in tightly-covered container in cool place.

Party Mixes

For any crowd set out the following:

#1 cashews (preferably unsalted) and raisins
#2 soy nuts or sunflower seeds, plain
#3 Roast in the oven: unsalted peanuts
coconut
honey
#4 almonds, dates, sesame seeds

Additional Sources

"How to Make Jams, Jellies, and Preserves at Home," a Department of Agriculture pamphlet from Public Works, edited by Walter Szykitka, Links Books, New York, NY, 1974.

The Complete Yogurt Cookbook, by Karen Cross Whyte, Troubadour Press, San Francisco, CA, 1970.
Making Homemade Cheeses and Butter, by Phyllis Hobson, Garden Way Publishers, Charlotte, VT, 1973.
Rose Recipes From Olden Times, by Eleanor Sinclair Rohde, Dover Publications, New York, NY, 1973.
Old Fashioned Recipe Book, by Carla Emery, Kendrick, ID, 1971.
Whole Earth Cookbook, by Sharon Cadwallader and Judi Ohr, Bantam Books, New York, NY, 1973.
Recipes for a Small Planet, by Ellen Buchman Ewald, Ballantine Books, New York, NY, 1973.

#36 DRINK HOMEMADE BEVERAGES

In one year the average American will consume about 35 gallons of soft drinks, 21 gallons of beer, 28 gallons of milk, 6 gallons of canned fruit juice, and 5 gallons of reconstituted and frozen citrus juice in addition to 13 pounds of coffee and 0.8 pounds of tea.[1] Also, many people consume far too much alcohol and coffee.

Since drinking water in America is fairly pure we are easily able to fill our daily fluid needs, yet we often desire something sweeter or more flavorful. The packaging and transportation energy required to bring non-nutritious soft drinks to our homes is a tremendous energy waste and makes processing more expensive than the actual value of the liquid contents. It takes over twice as much energy to manufacture a cola container than it does the contents! The disposal expense of these beverage containers is an additional cost and energy expenditure.

One way to curb container and transportation wastes is to make beverages at home. Apples, especially imperfect ones, can be used to make cider. A cider press for making juice is a simple device consisting of a chopper and a press, and can be used to make nutritious juice for home consumption.[2]

Fruits and berries can add body and flavor to fruit juices

and punch, while not being abundant enough to squeeze into pure juice (especially raspberries, blackberries, cherries, plums, and strawberries). Where grown in large quantities, some fruit such as peaches, pears, apricots, and grapes, are excellent blended with apple juice. When festive occasions deserve extra cheer, natural fermentation (a simple preservation method) can be used. Applejack can easily be made from a crop of apples.

Drinks can be made from grains and other vegetable products. Carrot and sauerkraut juice mix well with tomato juice and are rich in vitamins. Coffees and teas can be made from many products, including burned grains, herbs, rosehips, and cherry bark. All of these are money savers and eliminate caffeine from the diet. Spices such as cinnamon and cloves add flavor to teas. Lemon, lime, and orange pulp can add zest to teas and fruit punch, but do not use dyed citrus peels.

CARBONATED BEVERAGE ALTERNATIVES

- Cold water in a jar in the refrigerator.
- Freshly squeezed fruit juices.
- Canned and frozen concentrates of fruit and vegetable juices.
- Fresh milk or reconstituted dry milk.
- Punches made with fruit juice combinations.
- Fruit teas made from tea, strawberries, raspberries, lemons, oranges, cinnamon, cloves, honey, and other ingredients in various combinations.
- Apple cider or applejack.
- Home-brewed wine and beer.

Notes

1. "National Food Situation," Economic Research Service, U. S. Department of Agriculture, Washington, DC, February 1975.
2. *Making and Preserving Apple Cider,* USDA Farmers Bulletin #2125, Superintendent of Documents, U.S.G.P.O., Washington, DC 20402.

Additional Sources

Guide to Better Wine and Beer Making for Beginners, by S. M. Tritton, Dover Publications, New York, NY, 1970.

Nature's Drinks, by Shirley Ross, Vintage Books, New York, NY, 1974.

The Homemade Beer Book, by Vrest Orton, Charles E. Tuttle, Rutland, VT, 1973.

#37 ADVOCATE BREAST FEEDING

The best way to feed an infant is nature's way–breast feeding. The reasons are many: It is inexpensive and psychologically satisfying to both mother and baby. Physically, the baby benefits from the antibodies and enzymes in the milk, therefore, is protected from diseases from the earliest days of life. Mother's milk seems to be essential for infants to attain the proper metabolism of cholesterol in later life. Pediatricians believe that breast-fed babies have fewer allergy problems and are less likely to overfeed.[1]

HINTS FOR SUCCESSFUL BREAST FEEDING[2]

- Obtain information on breast feeding before the child's birth.
- Find a physician who supports breast feeding and insist on support from hospital staff.
- Get support from another nursing mother, friends, or an organization of women who nurse their babies.
- Resist promotions for formulas: ads and free samples from the hospital.

- Don't blame mother's milk for crying, spitting up, or unusual bowel movements.
- Persist.
- Get a good nursing bra that opens easily.
- Toughen the nipples with massage cream before the birth. Let baby nurse a few minutes the first day, gradually increasing the time. To prevent soreness allow 5 minutes on each breast. The baby gets four-fifths of the milk in the first five minutes.
- Touch baby's cheek with nipple to start. Allow baby to grasp the entire dark part of the breast in his/her mouth. When removing, break the suction by inserting a finger in the corner of baby's mouth.
- Wash nipples with mild soap and water at least once a day; rinse with clean water before and after nursing.

Unfortunately, breast feeding is not commonly practiced in this country, and when tried is stopped at the first sign of upsets. Even the developing countries are turning increasingly toward artificial feeding—a tragedy in these areas because the purified water and equipment needed for safe bottle feeding are not available. To help shift the public preference away from bottle feeding:

- The Department of Health, Education and Welfare should offer an educational program on the advantages of breast feeding.
- Public health clinics should promote breast feeding.
- Medical and nursing schools should devote more time to making the students more aware of the benefits and techniques of breast feeding.
- Hospital maternity personnel should encourage and assist nursing mothers.
- Employers should make time available so that working women can nurse their babies.
- The National Institutes of Health should further research the differences between breast and bottle feeding.
- The promotion of artificial feeding materials to other countries should be discouraged.

Babies do well on mother's milk exclusively for up to six months. Cereals introduced earlier can lead to digestive discomforts. Food companies often take advantage of inexperienced parents. Through advertising they convince the unsuspecting consumer that baby food in jars is the best, most modern way to feed babies. Grocery shelves are filled with jars of fruits, vegetables, meats, stews, cereals, and desserts. Fruits and vegetables cost an average of $.20 a jar; meat costs $.40 for one 3.5-ounce jar. Most baby foods can be made at home with very little effort and at half the cost.

Commercial baby food is often nutritionally inferior to home preparations. Sugar, modified starches, and water reduce the amount of real food, and other additives are not only unnecessary but may be harmful. Sodium nitrite, which may cause cancer, is sometimes used as a preservative in meat purées. Modified starch is added to prevent separation and has no nutritional value. Salt is sometimes added but is not needed since babies have no taste for salt. Water can be 70 to 90 per cent of the contents. After testing the major brands, Consumers Union found small insects, insect parts, rodent hairs, and paint chips (flakes from enamel on the insides of lids) in 25 per cent of the baby foods tested.

HINTS ON BABY FEEDING

- Use home-squeezed fresh orange juice.
- Feed the baby cereal when the doctor advises. When buying baby cereals, check the labels to avoid excessive additives and sugar. Use foods fortified with vitamins and iron.
- When baby is ready for other solid foods, purée fruits, meats, and vegetables at home. Use a blender, foodmill, grater, strainer, or fork, depending on the type of food. Make the purée as fine as is required by the baby's age and preferences.
- When cooking for baby use steam, a steam basket, or a double boiler. The vitamin loss is only 6 per cent, compared with a pressure cooker, 12 per cent, and boiling, 29 to 30 per cent. Do not add seasonings.

Notes

1. *White Paper on Infant Feeding Practices,* from CSPI Publications, 1757 S St., NW, Washington, DC 20009.

2. *Infant Care,* a booklet by the Office of Child Development, Superintendent of Documents, U.S.G.P.O., Washington, DC 20402.

Additional Sources

Womanly Art of Breast Feeding, LaLeche League International, Franklin Park, IL, 1963.

The Complete Guide to Preparing Baby Foods at Home, by Sue Castle, Doubleday & Co., Garden City, NY, 1973.

Infants and Mothers: Individual Differences in Development, by T. Berry Brazelton, Delacorte Press, distributed by Dial Press, New York, NY.

The Natural Baby Food Cookbook, by M. Kenda and P. Williams, Nash Publishing Corporation, New York, NY, 1972.

Make Your Own Baby Food, by D. and M. Turner, Workman Press, New York, NY, 1972.

#38 QUESTION PETS AND PET FOOD

Pets are the surrogate children, husbands and wives–of Western society, returning, for kibbles, kisses, companionship and devotion, or at least a cool tolerance accepted as love. Like the pharaohs, czars, and Caesars, Americans surround themselves with animals. In a disjointed society and a disquieting world, these anthropomorphized adoptees can be counted on to wag, purr, and warble, warming human hearts until they pass expensively on to await us in the Great Pet Sheraton Upstairs.

Reprinted by permission from *Time,*
The Weekly Magazine; Copyright Time Inc.[1]

Around the world, children go hungry and starve to death while millions of pets (100 million dogs and cats alone) consume billions of pounds of food, much of it high-protein meat. Americans spend $2.5 billion a year on commercially prepared pet food, which is equal to the dollar value of the food required to supply over one third of the world's hungry. Pets not only eat more nutritious meals than the world's poor, but they also receive vastly better medical care.

Veterinary hospitals and clinics in this country are almost as numerous as community health centers. In addition to medical care and grooming, an array of extravagant pet accessories and services are available, including: limousine services, psychiatric clinics, motels (equipped with color T.V. and air conditioning), beauty salons, pet jewelry, and even contact lenses.

The proliferation of pets (3,000 puppies and kittens are born every hour) has become a tax burden to citizens, as every day, dogs deposit an estimated 4 million tons of feces and 4.5 million quarts of urine on city streets.[2] Canine droppings are also a serious health hazard. Roundworm (toxocara) found in fecal matter can infect children playing in the street, and if left untreated can cause blindness.[3] Over 100 human infections, from tuberculosis to diphtheria, can be contracted by animals and passed on to humans. Exotic wild animals such as monkeys can carry rare diseases and remain dangerous to handle even after they are treated. Abandoned dogs that roam city streets are a serious safety problem. In New York City alone, 38,000 people were treated for dog bites in 1974.[4]

Some pets serve a necessary purpose in our society. Dogs can provide sight for the blind and protection from muggers and burglars. Many people depend on pets for companionship and affection, and children can learn about life and loving by owning a pet. If a child desires a new pet, consider that maybe a canary or goldfish could fulfill the need. They use far less of the world's limited food resources.

Anyone who raises a pet should do it properly. A responsible owner should check with a veterinarian about neutering the pet to prevent the proliferation of homeless animals. Persistent advertising by pet food manufacturers (who spend over

$165 million a year to advertise) has convinced many pet owners that canned meat is the best source of nutrition. Most cans are not even a good buy, since they often contain 75 per cent water. Dried cereals are more wholesome and contain only 10 per cent water. Domestic animals can be fed on table scraps supplemented with commercial dry cereals. Left-over meat scraps, trimmings, fish parts, and vegetable scraps can be frozen for later use as pet meals. All dogs can be healthy by eating a vegetarian diet. Help reduce the burden on the world's food resources by feeding pets vegetable protein.

Notes

1. "The Great American Animal Farm," *Time* magazine, December 23, 1974, p. 58.
2. Ibid., p. 59.
3. "Where Has My Little Dog Gone?" produced by Jim Jackson for CBS' "60 Minutes": CBS News, New York, NY, December 28, 1975.
4. *Time* magazine, *op. cit.*, p. 61.

Additional Source

Healthy Vegetarian Dogs and Cats, Juliette de Bairacli Levy, The Vegetarian Society, Cheshire, England.

IV GARDENING

For all things produced in the garden, whether of salads or fruits, a poor man will eat better that has one of his own, than a rich man that has none.

J. C. Loudon
An Encyclopedia of Gardening, 1826

INTRODUCTION

Many people have green thumbs but have become aware of it only in recent years. During World War II, 20 million "victory gardens" produced 40 per cent of the vegetables grown in this country. The 1970s have seen a renewed interest in home gardens, and the number increased to about 24 million in 1972, 27 million in 1973, and over 30 million in 1974, according to the National Garden Bureau of Gardenville, Pennsylvania.

Benefits in home gardening include:

- Economics: growing fresh produce will save money.
- Psychic health: working with soil and watching things grow is psychologically satisfying.
- Physical health: the exercise is good for us.
- Aesthetics: productive gardens are beautiful.
- Educational: a return to the soil broadens one's ecological interests, influencing the gardener to treat nature gently; it is an excellent way to teach children about nature.
- Conservation: the energy costs of transporting produce and fertilizing unproductive lawns are saved.
- Nutrition: vine-ripened produce, free from chemical pesticides, contains more nutrients and vitamins and fewer toxic substances.

#39 GARDEN ON AVAILABLE LAND

The amount of land in the United States used for growin
crops has been decreasing for more than twenty years. Al
though agricultural production has increased, we now plan
9 per cent less land in crops than in 1950.[1] This loss of agri
cultural land is most noticeable in urbanized areas wher
suburban sprawl, shopping centers, and highways have en
croached on farming communities. Many American resi
dences have front and back yards planted in lush, clippe
carpets of grass which require tender care, expense, fertilizer
and energy for lawn mowing. With an equivalent use of re
sources, productive gardens could blossom in our land instea
of manicured lawns. World food shortages should warn us t
make better use of our fertile soil. Unfortunately, many urba
areas are expanding into fertile agricultural areas. Becaus
green space is vital to good ecology, farms should not be re
placed with urban sprawl. Unfortunately, land use practice
(see entry #98) discourage farming near cities, and hig
taxes and real estate values pressure the few remaining farm
ers to leave the area or occupation.

Home gardening is a counter force to urbanization and

a good source of nutritious fresh produce. Vacant lots, abandoned right-of-ways, unused lawns, and river banks are ripe for urban gardening. Take precautions when gardening inner city plots. In heavily polluted urban areas, the soil could contain toxic metals such as lead, mercury, and cadmium. Contact the local or state agricultural extension station about these possible contaminants, and ask them what vegetables can be most safely grown in the soil. Lead could be a problem if plots are within 50 feet of a roadway. Incinerator locations are prime sites for metal contamination. In some instances, a 4–6-inch topsoil covering should first be applied. In others, leafy vegetables and tomatoes should replace root crops. Vegetables and fruits grown in heavily polluted areas should be washed thoroughly with vinegar and water to remove lead particles. City sludge can be a major source of fertilizer for urban gardens, but again some caution is due in heavily polluted areas because of metal contamination.

HELPFUL HINTS

- Ask homeowners to sacrifice some of their lawn and flower-bed areas to achieve a balance of food production and decoration. Vegetables such as tomatoes (regular and miniature), peppers, eggplants, lettuce, kale, and herbs are pleasing to the eye.[2]
- Use a mixture of grass, fruit trees, grapevines, and vegetables in neighborhood lots and plots.
- Establish a share program to facilitate recovery of vacant land and large unused yards. Find willing gardeners through local school, civic, and church organizations.
- Encourage growing vegetables and herbs in small plots and inside the home when land is unavailable. Beans, alfalfa, and other seeds can be sprouted indoors for food.
- Pressure the city and state government to inventory vacant land and to furnish gardening information to future gardeners.
- Initiate a school program to educate children in gardening. Perhaps use part of the school grounds for such a program. Extend the program to adults as well.

Notes

1. "Environmental Quality–1973," Fourth Report of the Council on Environmental Quality, Washington, DC, September 1973, p. 306.
2. "Mini-Gardens for Vegetables," House and Garden Bulletin #163, Superintendent of Documents, U.S.G.P.O., Washington, DC 20402.

Additional Sources

Gardening for Health and Nutrition, by Helen and John Philbrick, St. George Books, New York, NY, 1963.

A Modern Herbal, by M. Grieve, Dover Publications, New York, NY, 1967.

Organic Farming–Yearbook of Agriculture, by Ray Wolf, Rodale Press, Inc., Emmaus, PA 18049, 1975.

Farmers of Forty Centuries, by F. H. King, Rodale Press, Emmaus, PA 18049.

"Community Garden News," Suite G 17, American City Blvd., Columbia, MD 21044.

The Mountain Worker, edited by Chuck Smith and Sandy Adams, Catholic Worker Farm, Rt. 1, Box 308, West Hamlin, WV 25571.

Environmental Response
Box 1124, Washington University
St. Louis, MO

"Anti-Inflation Garden Program," J. B. Reagan
Pennsylvania Department of Agriculture
2301 N. Cameron St.
Harrisburg, PA 17110
(In its first year of operation this program provided low-priced seeds, educational literature, and even state-owned land for gardens to 200,000 people.)

#40 GROW VEGETABLES

High prices, chemically coated vegetables, artificially ripened produce, and additives in processed foods makes one look for ways to have wholesome food for the family. People are now using their yards, renting farm land, or sharing community gardens to have vine- or bush-ripened vegetables and fruits.

To obtain the maximum yield from invested time, labor, and money, learn the basics to avoid mistakes. Obtain information from the U.S.D.A. Agricultural Extension Service or garden club about the local soil structure and details on what vegetables grow best. Have a soil analysis to determine acidity and fertility.

Keep a record to avoid repeating errors of seed type, planting, and maturing times. Know the climate, water supply, and texture of the soil. Buy disease-resistant seeds from a reputable dealer. Generally, do not use home-grown seeds except for watermelon, muskmelon, and beans. Don't expect seeds to be viable for years. Corn, leeks, onions, parsley, rhubarb, and salsify seeds are good for only one or two years.

INDOOR AND HOTBED GERMINATION[1]

To get an early start, some seeds can be planted indoors and set outside when the climate permits.

- Six to ten weeks before outdoor planting, seed eggplant, tomatoes, and green peppers.

  Four to six weeks, seed broccoli, cabbage, cauliflower.
- Containers: Flower pots or flats, 3″ × 12″ × 18″.
- Soil: Equal parts of garden soil, compost, and builder's sand.
- Planting: Make a ½-inch furrow, water thoroughly, allow to drain. Sow seeds thinly in rows. Cover with a fine layer of sand or moss. Sprinkle with a fine mist.

Cover with a sheet of clear plastic film. (Plastic diffuses the light and holds the moisture.)

- Place in a south or east window, or in fluorescent lighting.
- Keep temperature at about 70° F.

When the first true leaves are formed, transplant the seedlings. Place each in a flower pot or milk carton (with holes in the bottom) or in large flats, allowing 2 inches per plant. Use a mixture of rich topsoil and compost. Moisten before moving, and handle with care to protect rootlets.

Two weeks before planting outdoors, harden the seedlings by withholding water and lowering the temperature. Ten days before planting, block the plants in flats by cutting through the roots around each plant.

Another way to start vegetable plants early in the spring is by using a hotbed. One can make a hotbed quite easily provided livestock manure is available. Use 2-inch lumber frames for support. An old wooden window sash can work well as the glass surface. It need not be hinged but can be manually removed from the frame on warm days. Build the plot on a slight slope, sloping toward south or west, to maximize sunlight and minimize drainage problems (see diagram).

WHEN TO PLANT OUTDOORS[2]

Vegetables differ in hardiness and in when they can be placed outdoors as seeds or small plants.

- Four to six weeks before frost-free dates: broccoli, cabbage, lettuce, onions, peas, potatoes, spinach, and turnips.
- Two to four weeks before frost-free time: beets, carrots, chard, mustard greens, parsnips, and radishes.
- On the frost-free date: snap-beans, okra, New Zealand spinach, soybeans, squash, sweet corn, and tomatoes.
- One week after frost-free date: lima beans, eggplants, peppers, sweet potatoes, cucumbers, and melons.

Transplant with plenty of dirt clinging to roots in the late afternoon. Place firmly in holes a foot or more apart. Pack soil firmly around plants and water well. If seeding, follow the directions on the seed packages.

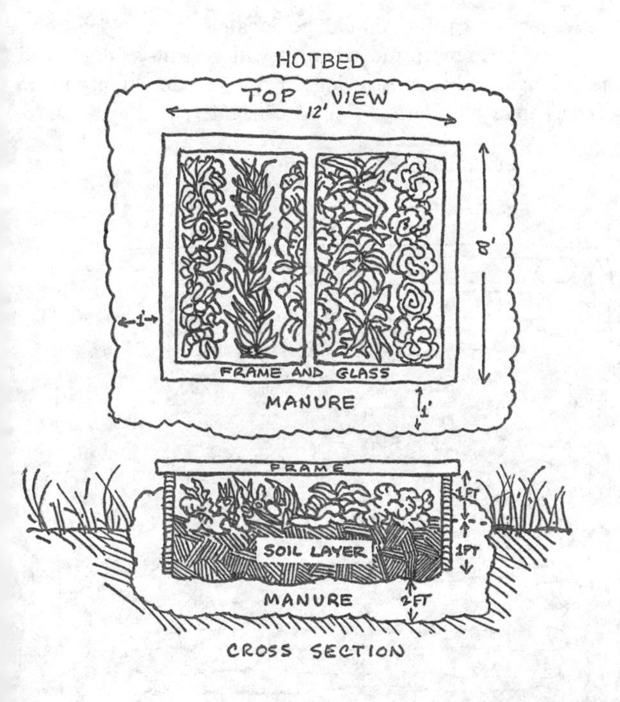
HOTBED
TOP VIEW
12'
8'
1'
FRAME AND GLASS
MANURE
1'
FRAME
1FT
SOIL LAYER
1FT
MANURE
2FT
CROSS SECTION

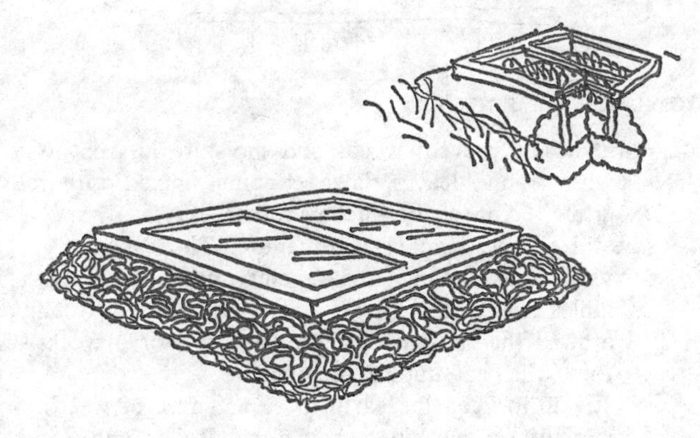

The vegetable garden should be located away from trees, protected with a fence, and where it will get full sunlight and good drainage. A bed measuring 25′ by 30′ can supply more vegetables than an average family can eat (see diagram).

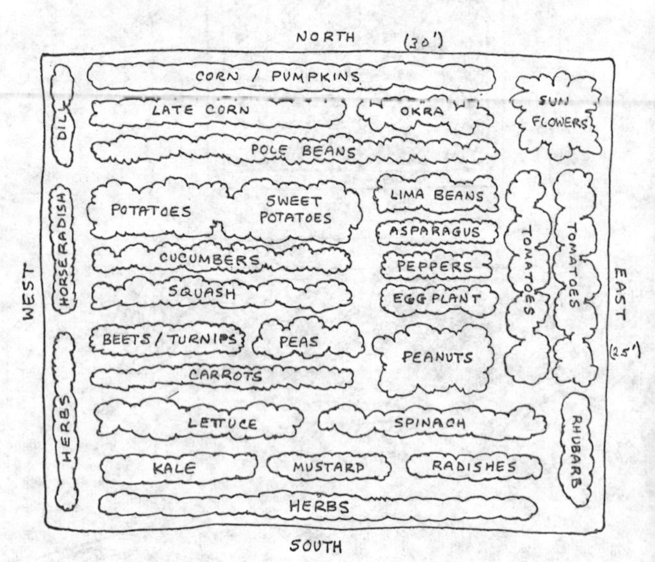

GARDEN CARE HINTS[3]

- Mulch: To prevent weeds and moisture loss cover beds with ground leaves, lawn clippings mixed with rotten straw, compost, or kitchen refuse.
- Fertilize: Use compost or manure. The cabbage group, vine plants, onion, tomato, celery, corn, and leafy vegetables need the richest fertilization. Put extra compost around the roots of tomatoes every 2 or 3 weeks and they will bear until September.
- Cultivate to keep the earth loose and free of weeds. Soil is right for working when a handful crumbles under slight pressure. Add more mulch when it has blown away or has been worked into the soil.

- Water when rainfall has fallen below one inch a week. A good soaking with one inch of water is better than many light sprinklings. Water in the evening to allow a night of soaking.

Grow peanuts along the south side of a wall; they need four months of warm weather to grow to maturity. Harvest before the fall frost. Dig up the whole plant and shake the soil from the roots (on which the nuts grow). Hang for three weeks in an airy, protected place. Remove the nuts and cure for two to three weeks in a cardboard carton. Roast in a shallow pan for 25 minutes to complete the process.

By planting hardy vegetables at the proper time you can eat garden produce through the fall and early winter. Plant part of the area with a slight slope, which prevents water saturated soil from freezing. Protect the plants from the cold with a heavy mulch cover. In temperate climates, turnips, kale, and mustard greens are ideal late crops.

Notes

1. *Growing Food the Natural Way,* by Ken and Pat Kraft, Doubleday & Co., Garden City, NY, 1973, pp. 145–99.
2. Ibid.
3. "Growing Vegetables in the Home Garden," Department of Agriculture Booklet, Superintendent of Documents, U.S.G.P.O., Washington, DC 20402.

Additional Sources

The New York Times Book of Vegetable Gardening, by Joan L. Faust, Quadrangle/Harper & Row, Scranton, PA, 1975.

How to Grow More Vegetables Than You Ever Thought Possible on Less Land Than You Can Imagine, by John Jeavons, Ecology Action, Palo Alto, CA 94306.

Ex-Urbanites Complete and Illustrated Easy Does It First Time Farmers Guide, by Bill Kaysing, Straight-Arrow/Simon and Schuster, New York, NY, 1975.

Grow It, by Richard Langer, Avon Books, New York, NY 10019, 1973.

How to Grow Fruits and Vegetables by the Organic Method, by J. I. Rodale, Rodale Press, Inc., Emmaus, PA 18049.

The Encyclopedia of Organic Gardening, edited by J. E. Rodale and staff, Rodale Press, Inc., Emmaus, PA 18049, 1959.

Farming with Nature, by Joseph A. Cocannouer, University of Oklahoma Press, Norman, OK.

Organic Vegetable Gardening, by Sam Ogden, Rodale Press, Emmaus, PA 18049.

Gardens for All
PO Box 2302
Norwalk, CT 06852

National Garden Bureau
Gardenville, PA 18926

#41 PLANT FRUIT TREES

Developing an appreciation for living things is beneficial to young and old. This can be done in the home environment by growing trees for fruit, nuts, shade, and aesthetics. Planting a tree, watching it grow, and harvesting the fruits can be an excellent family project. Perhaps each child could have a tree to care for. Mark an important occasion by a tree planting.

Trees are needed to provide oxygen and to absorb carbon dioxide from fossil fuel combustion and animal respiration. Arbors filter air, suppress noise, provide mulch, attract wildlife, and provide a substantial cooling effect.

FRUIT TREE VARIETY DIAGRAM

Fruit Tree Districts

1. *Apples, plums (artic and surprise).*

2. *Peach (arp-greensboro, halehaven), plum (imperial epineuse, stanley french), cherry (bing, windsor, reine hortense), apples, pears.*

3. *Peach (early crawford, belle, champion), plum (imperial epineuse, stanley french), cherry (bing, windsor, reine hortense), apples, pears.*

4. *Peach (early crawford, belle, champion).*

5. *Peach (early crawford, champion), cherry (bing), apples.*

Source: Garden Guide, *by Norman Taylor, Van Nostrand Reinhold: Cincinnati, OH, 1957, p. 484.*

Growing a tree is easy, but controlling disease and insects is more difficult. Confer with your local county agricultural extension agent when trees appear diseased. Know the characteristics of each species. A few features to look for are: tolerance to cold, size at maturity, fruit and flower characteristics, disease resistance, and soil requirements.

General rules for planting include: digging a large hole to allow the roots to spread; filling the hole with fertile soil containing humus or decayed plant matter; and watering generously after planting. Water frequently, at least one inch every week, especially during summer and early fall. Moisture deficiency will stunt growth and prevent proper root development.

Trees need sunshine for color and quality; therefore prevent root competition and lack of light by planting with an adequate distance between trees. To prevent crowding, plant with full tree maturity in mind. Prune occasionally to maintain shape and to keep main branches healthy.

When planting more than one tree, decide on the basic need–decoration, shade, windbreak, or orchard–keeping the total landscape in mind. Some trees grow to massive proportions and give excellent fall color, while others remain small yet produce brilliant and fragrant spring flowers. Some fruit trees bear in late spring while others produce in the early fall. The prospective planter should think ahead to the full-grown tree.

One of the easiest fruits to grow is apples. The tree withstands cold weather, resists disease, and responds well to pruning. Among the best varieties are Red Delicious, McIntosh, Jonathan, and Golden Delicious. Pears are another hardy fruit, with the Bartlett, Bosc, and Anjou the most common varieties. Peaches can be grown in most sections of the country. Early ripening varieties are best for eating fresh, while later-maturing varieties are best for home canning and freezing. Plum trees require care similar to that given peaches.

Other popular fruit trees that provide eating pleasure are citrus, figs, apricots, nectarines, and cherries. Check local nurseries and publications to determine which variety grows best in specific areas.

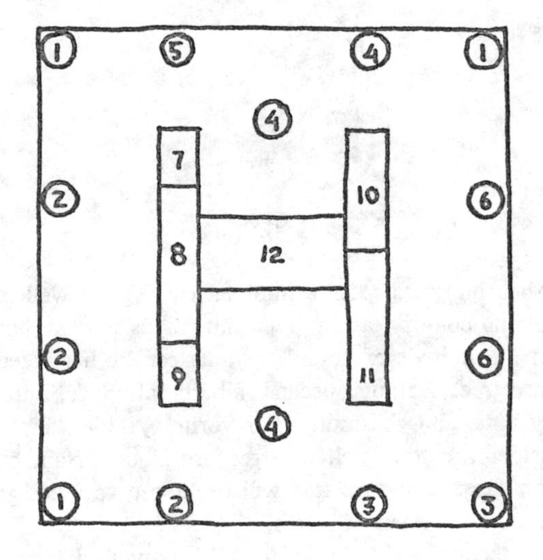

Fruit Garden 50 × 50 Feet

1. Apple–3 trees
2. Pear–3 trees
3. Cherry–2 trees
4. Peach–3 trees
5. Nectarine–1 tree
6. Plum–2 trees
7. Currant–4 plants
8. Blueberry–6 plants
9. Gooseberry–3 plants
10. Raspberry, 2 varieties–5 plants each
11. Blackberry, 1 variety–4 plants
12. Strawberry, 3 varieties–a double row of about 18 plants each, making 36 plants in all, depending on variety.

Standard Trees: (If your area is less than 50 × 50 feet, you must use dwarf or semi-dwarf trees.)

Source: Garden Guide, *by Norman Taylor, Van Nostrand Reinhold, Cincinnati, OH, 1957, p. 410.*

Nut-bearing trees require minimal care, grow well in most parts of the country, and provide nutritious food in bountiful amounts. Hundreds of pounds of nuts can be harvested from a mature tree. Walnuts, pecans, filberts, almonds, butternuts, hickory nuts, and chestnuts are favorite varieties. Two points to consider are how well the tree grows in certain climates and how large a mature tree will be. Some varieties grow to heights of more than 100 feet.

Local terrain and climatic conditions will determine the best arrangement of fruit trees in your garden or yard, but a general outline like the one on page 157 will be helpful.

References

The 1974 Yearbook of Agriculture–Shopper's Guide, by R. F. Carlson, Superintendent of Documents, U.S.G.P.O., Washington, DC 20402.

Teaching Organic Gardening, by the staff of *Organic Gardening and Farming Magazine,* Rodale Press, Inc., Emmaus, PA 18049, 1973.

Growing Food the Natural Way, by Ken and Pat Kraft, Doubleday & Co., Garden City, NY, 1973.

Additional Sources

Dwarf Fruit Trees for the Home Garden, by Lawrence Soutwick, Garden Way Publishers, Charlotte, VT 05445, 1972.

The Encyclopedia of Organic Gardening, edited by J. I. Rodale and staff, Rodale Press, Emmaus, PA 18049, 1959.

#42 FERTILIZE WISELY

The tremendous surge in this country's agricultural output during the last twenty-five years can be attributed in part to increased use of chemical fertilizers. These chemicals may benefit food production, but their widespread use has created an array of environmental side effects. With the chemical fertilizer demand growing at a rate of 9 per cent a year, the potential for environmental damage is also increasing.[1]

The environmental hazards associated with chemical fertilizer begin with its manufacture. Fertilizer requires tremendous quantities of natural gas. It is synthesized to produce ammonia and other nitrogen-based fertilizers. Domestic fertilizer consumption during the year ending June 30, 1974, was placed at 27 million tons. Of this amount, 4.2 million tons, or 15 per cent, is used on non-farm land, primarily on gardens and lawns.[2] This translates into an energy waste of 83,356 billion Btu's–waste–because means of natural fertilization are available to the grower that are not energy intensive.

Chemical fertilizers are fast-acting, short-term plant boosters that can seriously deplete the soil's growing capacity with extended use. Intensive application of chemical fertilizers, while giving short-term salutary effects, can create a demand for additional fertilizer in order to maintain balanced soil composition. Some soils accumulate residues of inorganic chemicals that lead to a relative decline in the humus content of the soil: The soil becomes less friable and less permeable to water and air. Continued deterioration creates hardpan, which hinders a plant's ability to absorb nutrients and increases runoff and erosion problems. Hardpans seal off the topsoil from the subsoil. Highly soluble chemicals, such as chlorides and sulfates, may also harm the beneficial microorganisms in the soil. Earthworms, essential soil cultivators, are destroyed in intensely fertilized areas. Finally, certain crops are more susceptible to disease if grown in excessively fertilized soil.

There are numerous pollution problems that result from the massive application of chemical fertilizers on cropland. Crops utilize only 50 to 60 per cent of the applied fertilizer.[3] The excess is leached away in runoff, which pollutes our lakes and streams, or is percolated downward into ground water. Well water in many parts of the country contains nitrate concentrations that pose a potential hazard to human health.[4] There is also growing concern with the nitrate levels in our food. During the breakdown of nitrates in the human digestive system, nitrites are formed that can form carcinogenic nitrosamines.[5]

HOW TO TEST SOIL

The most important part of gardening is to know your soil. If it is too wet, a drainage ditch may have to be dug. If it is deficient in the common nutrients–nitrogen, phosphorus and potash (N:P:K)–or some micro-nutrients, some provisions must be made to add them. This is done by testing the soil. Simple testing kits are available, but it may be wise to take a sample to a local county extension agent.

To take a sample of soil use clean tools and containers. Dig a 3″ core in 5 spots in your garden. Crumble the soil with an edged instrument and mix all 5 samples together on a flat surface. While still moist, put in a sample bottle, and cap.

The following list of natural fertilizers give many needed nutrients at little cost.

ALL-PURPOSE FERTILIZERS

Compost can include any organic waste that will decay in the soil. Compost is a storehouse for plant nutrients. For information on composting, see entry #44, "Compost and Mulch."

Leaves and Leaf Mold are an excellent source of humus, nitrogen, phosphorus, potassium, and mineral material, including calcium and magnesium. A good source of leaves is the closest vacant lot or woods.

Manure is a basic fertilizer that has been used for centuries. Horse, chicken, sheep, rabbit, hog, and cattle manure are in plentiful supply in many areas. If you live in the city, simply stop at a farm on a trip to the country and stock up on manure.

Sawdust and Wood Chips provide essential nutrients, aerate the soil, and increase its moisture-holding ability.

Activated Sludge is produced when sewage is agitated by bubbling air through it. Digested sludge, formed when sewage is allowed to settle over filter beds, has a nutrient value similar to manure. Sludge can be obtained from local sewage treatment plants.

Hay, Straw, and Peat Moss do not contain large quantities of nutrients, but they do aerate the soil, improve drainage, and aid plants in absorbing nutrients from other materials.

NITROGEN SOURCES

Blood Meal and Dried Blood can be obtained from a slaughter house, garden store, or fertilizer supplier. Because of its high nitrogen content, a sprinkling is enough to stimulate bacteria growth in the soil. Many gardeners add dried blood to water and spray compost heaps to hasten plant decay.

Grass Clippings are a rich source of nitrogen when worked into the soil or added to compost heaps.

Fish Meal and Fish Scraps are fertilizer sources common to any coastal or lake region.

PHOSPHATE SOURCES

Phosphate Rock and Colloidal Phosphate are excellent sources of phosphorus that also contain valuable trace elements, including calcium, iron, magnesium, sodium, iodine, and boron. These fertilizers are available in garden shops and fertilizer supply houses.

POTASSIUM (POTASH) SOURCES

Green Manure such as clover and alfalfa when plowed under is a useful source of potassium and nitrogen.

Seaweed and Kelp can be used as mulch, worked directly into the soil, or added to compost. Seaweed is high in potash (about 5 per cent) and is available in any coastal region.

Wood Ashes contain at least 7 per cent potash and are alkaline. They can be mixed easily with other fertilizing materials to make a quality fertilizer.

When mixing fertilizer, combine as many of the preceding materials as possible, thus creating a complementary amalgamation. The objective of organic fertilization is to feed the soil, not just supply the nutrients needed to produce a single crop in one season. Most organic fertilizers are slow-acting; therefore, the organic materials added to the soil have a residual effect that improves the soil for many years to come.

Notes

1. "Commercial Fertilizers: Consumption in the U.S. Year Ended June 30, 1974," CROP Reporting Board, U.S.D.A., Superintendent of Documents, U.S.G.P.O., Washington, DC 20402, November 1, 1974, p. 3.
2. *Energy and Food,* by Albert J. Fritsch, Linda W. Dujack, and Douglas A. Jimerson, CSPI Reports, 1757 S St., NW, Washington, DC 20009, 1975, p. 8.
3. "Fertilizer Use and Water Quality," by G. Stanford, Agricultural Research Service, U.S.D.A. Publication, Superintendent of Documents, U.S.G.P.O., Washington, DC 20402, 1970.
4. "Poisoning the Wells," Staff Report, *Environment,* January/February, 1969.
5. *Nutritional and Environmental Aspects of Organically-Grown Food,* by Mark Schwartz, Rodale Press, Emmaus, PA 18049.

Additional Sources

Teaching Organic Gardening, by Rita Reemer, Rodale Press, Inc., Emmaus, PA 18049, 1973.

Growing Food the Natural Way, by Pat and Ken Kraft, Doubleday & Co., Garden City, NY, 1973.

Organic Fertilizers: Which Ones to Use and How to Use Them, by staff of Organic Gardening and Farming, Rodale Press, Inc., Emmaus, PA 18049, 1973.

#43 USE NATURAL PESTICIDES

The use of pesticides and herbicides in this country tragically demonstrates the paradox of twentieth-century technological progress. These "miracle" chemicals (over 60,000 pesticide preparations alone)[1] improve agricultural yields by controlling insects and pests, make lawns beautifully green, and impart considerable comfort during the summer mosquito season. However, these same wonder chemicals that have brought about these advances have also created catastrophic ecological disorders. Pesticides have inobtrusively infiltrated every natural ecosystem, silently killing and contaminating animal species (through the food they eat) from tiny insects to man himself. One need only point out adverse environmental effects induced by DDT. Rachel Carson's predictions of the early 1960s about explosive environmental damage are now being borne out.

Because of the relative abundance of fossil fuels, chemical pesticides and herbicides have been inexpensive. However, it takes massive amounts of energy to produce the pesticides that are sprayed on our croplands, livestock, and residential land areas. Almost 400,000 tons of pesticides are used on major crop and livestock products in this country. It requires 26.5 trillion Btu's to produce these pesticides.[2] Such wastefulness can be avoided by utilizing nature's own methods of pest prevention. To promote natural pest control for large-scale farming is a massive endeavor. The following alternatives lend themselves to family gardening practices.

Diversity is the key to nature's stability. Pesticides destroy this stability by unselectively killing many insects and non-pests that co-exist with target species. When a pesticide is applied, it creates an imbalance in the ecosystem, and allows other pest species, once held in check by a predator or parasite, to multiply and create additional problems. The resistant mutants in target species survive and multiply during extended pesticide use and necessitate repeated applications of the same chemical at higher concentrations or new applications of different chemicals.

With natural methods of control organic gardeners are able to eliminate troublesome pests without large expenditures of time, money, or energy. One pesticide application for a specific pest may overturn the whole environment of a garden. Therefore, it is much better to rely on natural methods of control.

Preventing trouble is always better than having to cure it. The following are ways to prevent problems:

- Buy disease- and pest-resistant plant varieties.
- Plant at the correct time–insect life cycles vary; one may be able to avoid the pest problem altogether.
- Choose the site carefully–locate the garden where it can get adequate sunlight, air circulation, and water drainage.
- Keep the garden weeded and watered.
- Spot planting–growing a type of plant in a few sections of the garden can prevent complete destruction if pests do attack one part of the garden.
- Turn over the garden in the late fall–if the garden is troubled by pests that live in the soil during cold weather, spade in the fall and leave the earth exposed.
- Rotate crops–shift growing locations of vegetables; this practice inhibits the stabilization of pest populations.
- Mixed or companion plantings are effective methods for controlling pest populations. Plants can affect those they are growing near by helping or hurting them. Herbs, when used as borders, are effective in repelling many insect varieties. Leeks help carrots, and potatoes make an excellent companion for corn or beans. Nasturtiums planted near fruit trees discourage aphids. Tomatoes with asparagus acts against the asparagus beetle. Green

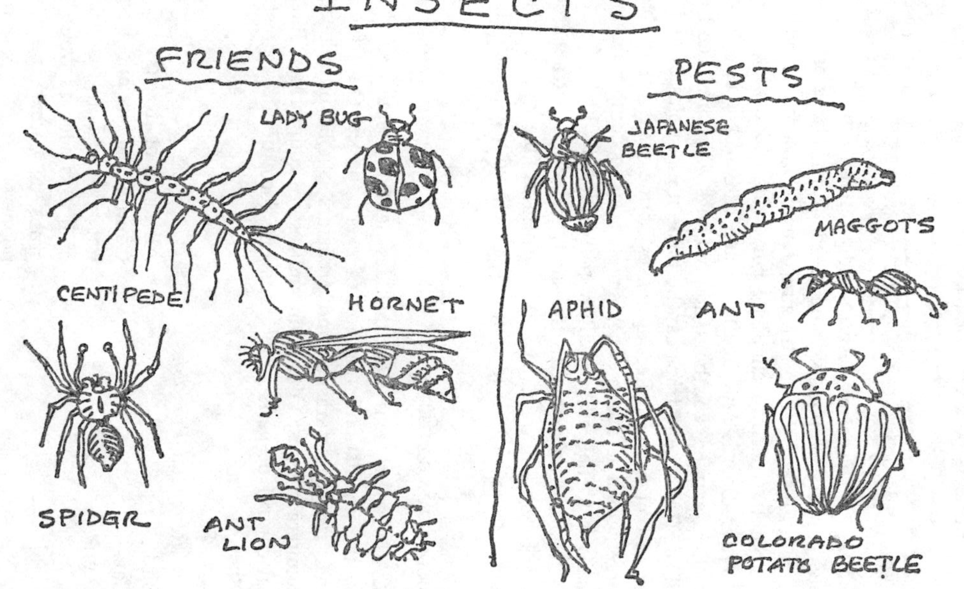
INSECTS
FRIENDS
PESTS
LADY BUG
CENTIPEDE
HORNET
SPIDER
ANT LION
JAPANESE BEETLE
MAGGOTS
APHID
ANT
COLORADO POTATO BEETLE

beans with potatoes prevents infestation of the Mexican beetle and the Colorado potato beetle. Horseradish protects potatoes from potato bugs.

MECHANICAL CONTROL OF INSECTS AND PESTS

Fences—The term "pests" usually refers to insects, but rabbits, gophers, chipmunks, and raccoons can also create havoc in the garden. When stringing fence, be sure to bury the bottom 10″ deep to eliminate burrowing rodents.

Traps—Boards laid in the garden can attract insects such as slugs, earwigs, sow bugs, and pill bugs which are then easily killed. Homemade traps can be made using molasses and water as bait or by collecting leaves of the plant to which the target insect is attracted. A simple way to control slugs and snails is to half fill a shallow pan with stale beer. The beer itself is not poisonous to the snails and slugs, but apparently they become intoxicated and drown in it! One way to protect plants from birds is to spot-cover them with cheesecloth.

BIOLOGICAL CONTROLS

A balanced ecosystem includes natural enemies as checks on the population expansion of every species. Once the balance is upset, parasites and predators can be introduced to counteract a pest infestation and restore equilibrium. Biological control of pests simply sets one form of life against another.[3]

Beneficial Predators	*Pests*
Bird	All kinds of insects
Cat	Controls rabbits and mice
Skunk	Eat mice and a great variety of insects such as white grubs, cutworms, and grasshoppers
Shrews	Mice and insects
Toads and lizards	Controls insects
Weasels	Voracious eater of rats and mice

Beneficial Insects	*Target Pests*
Aleochara beetles	Maggots
Ant lions	Ants
Aphid lions	Aphids and other harmful insects
Assassin bugs	Feed on insect larvae
Centipedes	Slugs and larval insects
Dragonflies	Catch flying insects, especially mosquitoes
Ground beetles	Caterpillars are its main diet
Hornets and yellow jackets	Keep down caterpillars and flies
Ladybugs	Aphids, spider mites, scales, and mealy bugs
Pirate bugs	Mites and eggs and larvae of many insects
Praying mantis	A predator that eats almost everything
Spiders	A variety of insects
Wasps	Caterpillars and other insects

HOMEMADE, ENVIRONMENTALLY SAFE, NATURAL PESTICIDES

Soap spray (1 cup soap flakes with 1 gallon water) is particularly effective against aphids.
Sulphur powder, sprinkle on potato slices prior to planting.
Wood ashes sprinkled around stems deters maggots.

Mineral oil will kill corn-ear worms if applied after the silks go limp and their ends start to brown.

Skim milk spray has been proven effective in controlling mosaic, which affects peppers and tomatoes.

Table salt spray (2 teaspoons to 1 gallon of water) can be used against cabbage worms and some mites.

Tomato leaf spray (made by steeping tomato leaves in water) is mildly effective against tomato plant pests.

HERBICIDES

Hand pull weeds instead of using herbicides, or mulch generously to prevent their growth. Cover garden with plastic in the fall to prevent weed seed germination.

For information on biological control of pests:

International Center for Biological Control
University of California
1050 San Pablo Ave.
Albany, CA 94706

This organization offers technical information on the natural enemies of particular pests. They also offer advice and counsel on biological control matters.

Notes

1. *The Complete Ecology Factbook,* by Philip Nobile and John Deedy, Anchor Books, New York, NY 10017, 1972, p. 293.
2. *Energy and Food,* by Albert J. Fritsch, Linda Dujack, and Douglas Jimerson, CSPI Reports, 1757 S St., NW, Washington, DC 20009, 1975, p. 7.
3. *Growing Food the Natural Way,* by Pat and Ken Kraft, Doubleday & Co., Garden City, NY, 1973, pp. 113–16.

Additional Sources

"The Living Garden," Audubon Naturalist Society, Washington, DC 20015.

Teaching Organic Gardening, by Rita Reemer, Rodale Press, Emmaus, PA 18049, 1973.

The Organic Way to Plant Protection, by Organic Gardening staff, Rodale Press, Emmaus, PA 18049.

Organic Gardening Without Poisons, by Hamilton Tyler, Van Nostrand Reinhold Books, New York, NY 10018, 1970.

Grow It Safely, by Stephanie Harris, Health Research Group, Washington, DC 20036, 1975.

Companion Plants and How To Use Them, by Helen Philbrick and Richard G. Gregg, Devin-Adair Co., New York, NY, 1966.

The Bag Book, by Helen and John Philbrick, Garden Way Publishers, Charlotte, VT 05445, 1974.

Silent Spring, by Rachel Carson, Houghton-Mifflin, New York, NY, 1962.

#44 COMPOST AND MULCH

The United States generates about 40 million tons of garbage each year.[1] This includes uneaten food, spoiled food, food scraps, and yard wastes. Only an affluent society like ours can afford to throw away this much food. Third and Fourth World peoples produce little or no garbage, for whatever is not eaten is fed to animals or returned to the soil.

When America was primarily a rural nation, there was no garbage problem. But as cities grew, garbage disposal became a problem, and open burning dumps appeared in every community. The garbage dump became a serious health threat, and clouds of smoke polluted the air. Most city dumps were replaced by enclosed incinerators. But air pollution remains, and incineration wastes valuable organic energy sources.

Cities are still wrestling with the garbage-disposal problem. Many bury wastes in landfills. However, land for this purpose is becoming less available, the cost of transporting the wastes is prohibitive and runoff from these sites poison surface and ground water supplies. Kitchen garbage disposals once seemed to provide the answer, but these expensive appliances require water and electricity. Local sewage-treatment plants are drowning in liquid sewage, and a valuable fertilizer source is being lost.

One solution to our disposal difficulties lies in conserving nutrients in food wastes by returning them to the soil either by direct burial or in the form of compost. Put kitchen scraps into milk cartons, and when full, place them in a compost pile. Soggy food wrappings, soiled egg cartons, and other waste paper not suitable for recycling can be shredded and added to the compost with garden cuttings, leaves, and degradable household wastes and used as fertilizer in the garden.

Methods of composting include:

ODORLESS KITCHEN COMPOST FOR APARTMENTS[2]

1. Cut off the side of a large milk carton and staple the top shut.
2. Cover the bottom with a half inch of soil.
3. Dice wet garbage and stir into the soil.
4. Sprinkle with water to keep moist.
5. Stir several times a day.
6. Once a week add more soil.
7. Keep moist and stirred and the vegetable matter will not give off any odors.
8. When full, use the compost to pot a plant or put it into your garden and start another batch.

METAL COMPOST UNIT

An enclosed metal rodent-proof composter can be made from a 55-gallon oil drum. The top has a lid to keep animals out while the bottom is removed so that water from the decomposition process can drain into the soil. Also, microorganisms in the soil can encourage decomposition.

The top of the barrel is large enough to allow removal of the compost by shovel. Meat fat and bones will lose their odor in this unit and can be added to gardens without attracting dogs or other animals.

The lid is attached to the barrel with a cross section of rubber tire, which serves as a flexible hinge; the tire is bolted to the lid and barrel. A metal hasp is attached to the front of the barrel to keep raccoons out. The barrel and hardware are painted with rust-proof paint.

For more information contact:

Martha Stone
42 Wachusett Rd.
Wellesley, MA 02181

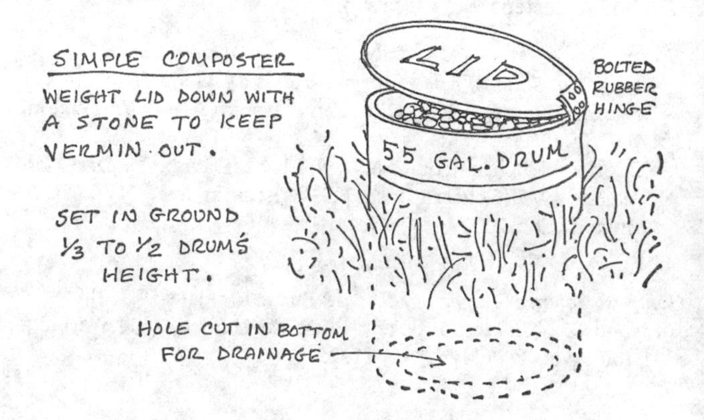

Mulch is a layer of material applied to the soil surface to protect plant roots from excessive heat, cold, or dryness. It discourages the growth of weeds and helps maintain good soil structure by preventing the soil from crusting and compacting. The amount of water evaporating from the soil can be reduced by 10 to 50 per cent by mulching. The decomposed matter supplements the organic content of the soil, enriches the soil structure and makes it more productive.

MULCHING MATERIALS

- Solid wastes from the house:
  Composted kitchen scraps—vegetable trimmings, inedible left-overs, moldy or rotten fruits and vegetables.
  Degradable wrappings—all kinds of shredded food wrappings.
  Dirt and dust collected by cleaning.
- Garden leavings: straw, grass cuttings, partially decayed leaves, sawdust, wood chips, peanut hulls, twigs, and pine needles.
- Sheets of newspaper or black plastic: Cover the bed with a large piece of black plastic; hold the edges with earth; make slits for the plants and punch holes to let water in. Black plastic acts like a greenhouse, retains heat, and hastens early growth.

Notes

1. *Energy in Solid Waste,* Citizen's Advisory Committee on Environmental Quality, 1700 Pennsylvania Ave., Washington, DC 20006.
2. "Odorless Kitchen Compost," by Elaine Davis, *Concern Calendar, 1975,* Concern, Inc., 2233 Wisconsin Ave., NW, Washington, DC 20007.

Additional Sources

Compost Science Magazine, Rodale Press, Emmaus, PA 18049.
Handbook on Mulches, Brooklyn Botanical Garden, 1000 Washington Ave., Brooklyn, NY 11225.

#45 KEEP BEES

Beekeeping is a novel hobby that is attracting a number of urban and suburban natural food enthusiasts. "Homemade" honey is not only cheaper than store-bought honey, but is often more flavorful. As a natural sweetener, honey can be used in place of refined sugar.

Requirements for honey production are:

- Hives, frames, and a foundation to support honey combs.
- Hive tools to pry apart the frames.
- A smoker, to blow smoke into the hives to pacify the bees when working with them.
- A veil and gloves for protection.
- Two or three pounds of bees, and a queen.

Hives can be purchased from a factory or built at home (see diagram).[1]

Start the hive in the spring when fruit and flower blossoms are available for nectar. Hives should be placed near a fresh-water supply and away from where people might be stung. In warm areas, pick a shady spot. Where temperatures fall to freezing, put hives in the sun and protect them from wind.

Many nectar sources are available: alfalfa, aster, buckwheat, catclaw, citrus fruit, clover, cotton, firewood, goldenrod, holly, horsemint, mesquite, palmetto, tulip tree, tupelo, sage, sourwood, star thistle, sweet clover, sumac, and willow. Depending on the nectar source, the honey may be colorless, amber, or red; the flavor will vary from mild to strong.

To start a new hive, put a syrup mixture of half molasses and half water in a feeder at the entrance. As the colony grows, make more room by adding extra boxes of combs to the hive.

When removing honey from the hive, work in good weather, when the bees are actively flying. Wear protective clothing: a veil over the head and face; close-woven, light-colored clothing, sealed at the ankles and wrists; and gloves. Before disturbing the bees, direct a smoker into the hive entrance. When removing the cover, gently spray smoke at the exposed bees. If stung, quickly remove the stinger by scraping it with a fingernail. After a few seasons, many beekeepers develop immunity to stings.

When removing honey leave plenty for the bees; a typical hive needs about 50 pounds to survive the winter.

For a good beekeeping catalog and a book of instructions, write:

Walter T. Kelley Bee Supply Market
Clarkson, KY 42726

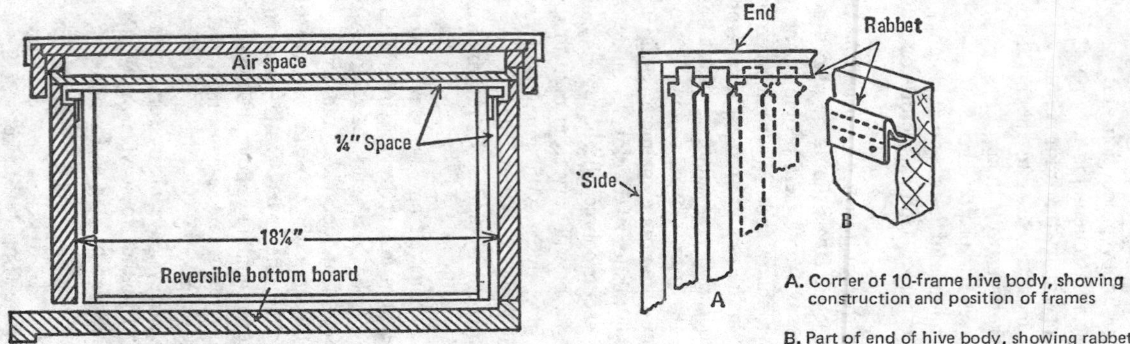

CROSS SECTION OF HIVE BODY AND FRAME

A. Correr of 10-frame hive body, showing construction and position of frames

B. Part of end of hive body, showing rabbet, which should be made of tin or galvanized iron

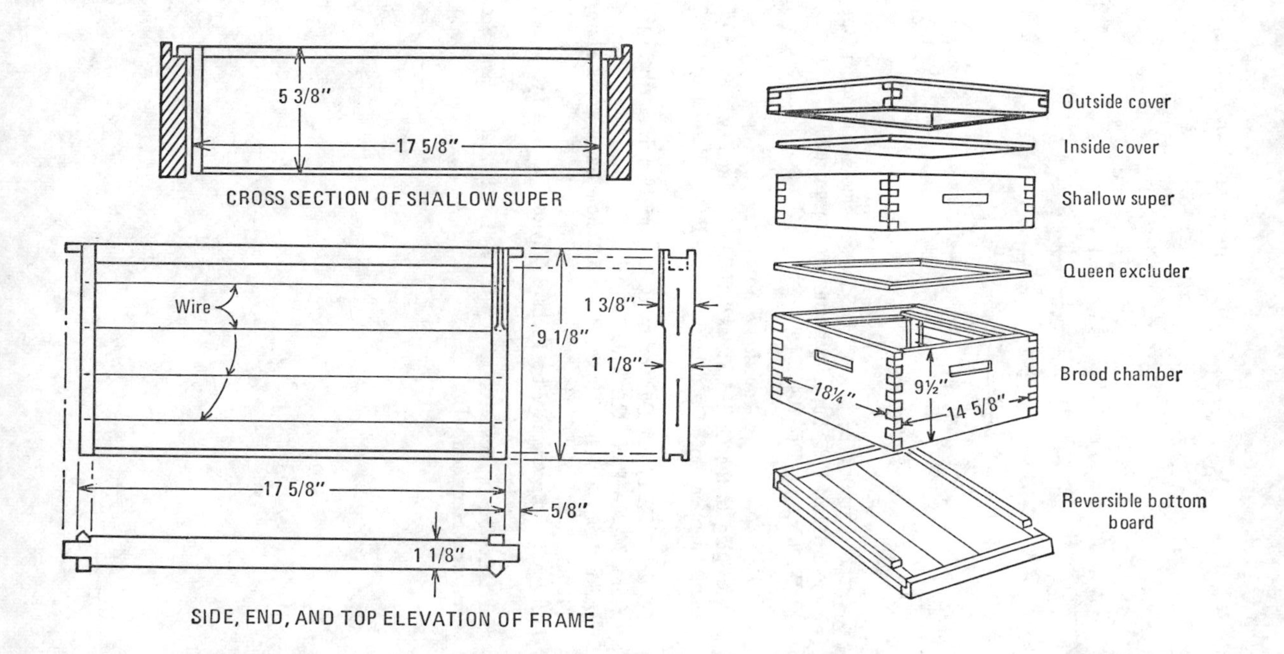
5 3/8″
17 5/8″
CROSS SECTION OF SHALLOW SUPER
Wire
9 1/8″
1 3/8″
1 1/8″
17 5/8″
5/8″
1 1/8″
SIDE, END, AND TOP ELEVATION OF FRAME
Outside cover
Inside cover
Shallow super
Queen excluder
Brood chamber
18¼″
9½″
14 5/8″
Reversible bottom
board

Note

1. "Beekeeping for Beginners," U.S.D.A. Home and Garden Bulletin No. 158, Superintendent of Documents, U.S.G.P.O., Washington, DC 20402.

Additional Sources

"Beekeeping in the United States," U.S.D.A. Handbook No. 335, Superintendent of Documents, U.S.G.P.O., Washington, DC 20402.

ABC and XYZ of Bee Culture, by E. R. Root, Whole Earth Truck Store, Menlo Park, CA 94053.

Starting Right With Bees, by A. F. Root, Whole Earth Truck Store, Menlo Park, CA 94053.

The Joys of Beekeeping, by Richard Taylor, St. Martin's Press, 1974.

"First Lesson in Beekeeping," by C. P. Dadant, *American Bee Journal,* Hamilton, IL.

V SOLID WASTE

INTRODUCTION

When thrift was an essential part of American life, solid waste disposal was no problem. Everything was used and re-used. We generate so much solid waste that collection and disposal will cost the nation almost $45 billion between 1973 and 1982.[1] Disposing of glass, plastic, paper and metal products is now one of the costliest items in a city's budget.

More than 125 million tons of household, commercial, and institutional (schools and hospitals, for example) solid waste is generated annually in America.[2] Although this figure excludes industrial, agricultural, animal, and mineral wastes, the total amount that ends up in the National Trash Can is tremendous (see diagram). In effect, each person produces 3.2 pounds of solid waste a day. The waste of energy represented by this mountain of trash is considerable. If the 80 per cent of combustible trash was burned as fuel, it would theoretically represent the equivalent of 8 billion gallons of gasoline or 178 million barrels of crude oil.[3]

Since few municipalities have policies dealing with waste generation, individual action is required. Do not purchase disposables; sort out junk items; use recycling centers; compost and mulch. Try reducing household waste by half—then by half again. One may find that so much trash is reusable, biodegradable, or recyclable that the sanitation department will have little to collect.

Notes

1. "Calculating Abatement Costs," Council on Environmental

NATIONAL TRASH CAN
(Annual Municipal Solid Waste—1971)

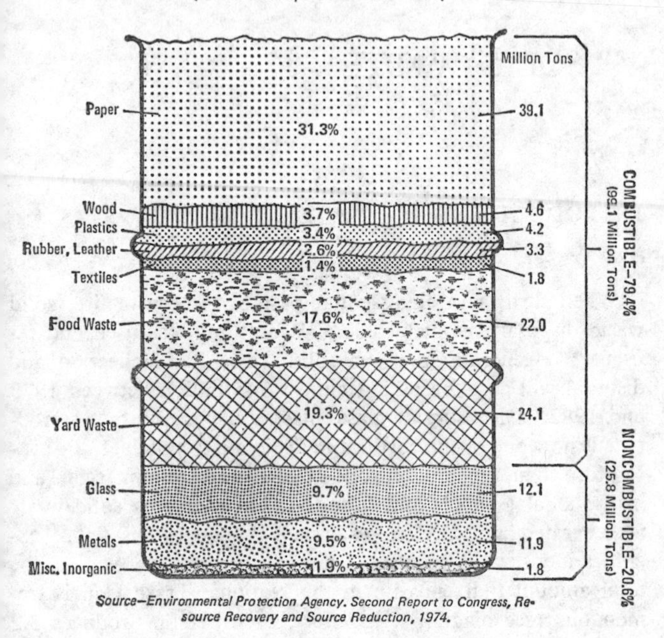

Source—Environmental Protection Agency. Second Report to Congress, Resource Recovery and Source Reduction, 1974.

Quality Fifth Annual Report, Superintendent of Documents, U.S.G.P.O., Washington, DC 20402, 1974, p. 224.

2. *Energy in Solid Waste,* Citizen's Advisory Committee on Environmental Quality, 1700 Pennsylvania Ave., Washington, DC 20006, p. 4.

3. Ibid.

#46 AVOID DISPOSABLE PAPER PRODUCTS

Disposable paper products represent a tremendous waste of raw materials (trees) and energy. American per capita consumption of paper tissues and towels is almost 860,000 Btu's each year. To prevent such waste, do not purchase disposable paper items.

Avoid eating at fast-food chains that use excessive packaging. When shopping in the supermarket, do not buy overpackaged foods and beverages.

Use items that can be used again:

- A handkerchief instead of paper tissues.
- Cloth or a torn sheet instead of facial tissue.
- Cotton towels or rags instead of paper towels. Every household has plenty of worn clothing that could be used for cleaning.
- Cloth napkins instead of paper ones.
- China, glass, and ceramics instead of paper plates and cups.
- Washable glasses in the bathroom instead of paper ones.
- Toilet paper must be used, but purchase only white paper; the dyes in colored paper pollute the water.
- Disposable diapers foul household plumbing, add to our waste of resources, have a non-biodegradable plastic covering, and can carry disease when placed in the trash. It's an expensive way to care for the baby. Comparative prices show the average yearly cost of diapers: cloth washed at home–$60; cloth washed at laundromat–$80; cloth diaper service–$170; paper diapers–$240.[1]

• Discourage junk mail by returning it to:

Consumer Relations Manager
Direct Mail Advertising Association
230 Park Ave. South
New York, NY 10017

Note

1. "Disposable Diapers," *Consumer Reports,* February 1975, pp. 98–100.

References

Garbage Guide, Environmental Action Foundation: Dupont Circle Bldg., Suite 724, Washington, DC 20036.

Reduce–Targets, Means and Impacts of Source Reduction, from the League of Women Voters of the U.S., 1730 M St., NW, Washington, DC 20036.

#47 REFRAIN FROM PURCHASING PLASTIC-WRAPPED ITEMS

During 1971 the consumption of plastic packaging in the United States was 2.5 million tons[1] (out of 10 million tons of plastic produced).[2] This is equivalent to 25 pounds per person. By 1980 industry officials expect these figures to triple. Plastic packaging litters the environment and creates serious solid-waste management problems.

When buried as landfill, plastic does not decompose through bacterial decay. If burned, plastic leaves a non-biodegradable residue and produces toxic air pollutants such as carbon monoxide and hydrochloric acid. These non-degradable properties prevent the recycling of the petrochemicals that compose the plastic. To make one ton of plastic, 72 gallons of refined crude oil, 338 gallons of natural gas liquids, and 37.1 million Btu's of energy in production are required.[3] This represents a total energy cost, fossil fuel for production and petrochemicals in the plastic, equal to 10,500 billion Btu's per year (1971). If projected increases in the production of plastics are met by 1980, this would mean an energy cost equal to 31,500 billion Btu's.

The Food and Drug Administration has proposed banning polyvinyl chloride plastic food containers (50 million lbs. per year in the U.S.),[4] as a result of a petition filed by Ralph Nader's Health Research Group indicating that vinyl chloride, the chemical used in the production of polyvinyl chloride plastics, can cause a rare type of liver cancer, angiosarcoma.

The adverse health consequences result from vinyl chloride, a known carcinogen, leaching into the food from the plastic containers. The critical nature of this threat becomes apparent when one considers the vast number of supermarket and household items packaged in polyvinyl chloride (PVC) plastic coatings and containers. From meat to milk, and bleach to mouthwash, PVC plastic containers are widely used for packaging. Because plastics create environmental pollution problems, deplete our dwindling natural resources, and pose serious health concerns, it is recommended that consumers refrain from buying plastic wrapped items, especially plastic food packaging.

Helpful Hints

- Reuse plastic bags already accumulated for non-edible materials and garbage.
- Buy food in bulk; it requires less packaging. Insist on paper rather than plastic wrapping.
- Shop at food co-ops and buy fresh produce that requires no packaging.
- Use old glass containers to store food.
- Complain to grocers if there is no alternative to plastic food packaging.
- Write to legislators to change the laws regarding plastics, particularly those involving food packaging.

Notes

1. *Energy in Solid Waste: A Citizen's Guide to Saving,* Citizens Advisory Committee on Environmental Quality, 1700 Pennsylvania Ave., Washington, DC 20036, p. 3.
2. "Bureau of the Census, Tariff Commission," *Chemical and Engineering News,* December 24, 1973, p. 8.
3. "The Environmental Impacts of Packaging," by E. L. Claussen, Office of Solid Waste Management, U.S.E.P.A., Washington, DC, 1973, pp. 4–6.
4. "PVC Makers Confident on Food-Contact Uses," *Chemical and Engineering News,* September 15, 1975.

Additional Sources

Health Research Group
2000 P St., NW
Washington, DC 20036

Environmental Educators, Inc.
1621 Connecticut Ave., NW
Washington, DC 20009

Garbage Guide, Environmental Action Foundation, Dupont Circle Bldg., Suite 724, Washington, DC 20036.

#48 BAN THE NON-RETURNABLE

In 1974 almost 32 billion cans and more than 15 billion disposable bottles were sold in America.[1] They litter parks, streets, waterways—just about everywhere. Most throwaways end up in the city dump or in sanitary landfills, which indicates confidence that the world contains enough space to endlessly receive junk. This approach assumes that the materials and energy for making these containers are inexhaustible.

A glass container made from silica sand is not a serious drain on natural resources; but an aluminum can uses 4 tons of bauxite per ton of aluminum and is a costly drain on nonrenewable resources.

Producing a 12-ounce throwaway can consumes 2.9 times

more energy than a 12-ounce returnable bottle (presuming 15 returns per bottle). Reusing glass bottles is the most energy and resource saving way to market beverages. Deposits paid when the drinks are purchased and refunded when the bottles are returned, are the best assurance that the bottles will be returned. Each refill from a returnable bottle represents one container that does not end up in the waste disposal system.

The Seven Sins of Throwaways[2]

1. Massive energy waste
2. Squandering the earth's resources
3. Contribution to environmental degradation
4. Economic loss to consumer and nation
5. Addition of solid waste
6. Employment loss
7. Litter

Oregon passed a Minimum Deposit Act in 1972. The minimum deposit is 5 cents for all cans and all brand identifiable bottles, and 2 cents for bottles that can be refilled by more than one bottler. Despite industry's resistance and predictions of economic disaster, the law has been successful. Throwaway glass containers have been virtually eliminated; only 1 per cent of all soft drinks and 0.05 per cent of all beer were sold in cans in 1973. Not one brewer, soft-drink bottler, or distributor has gone out of business. Litter has been drastically reduced and the total amount of solid waste decreased.[3]

Back citizen groups at local, city, and state levels and the legislators who are working for bills against the non-returnables.

Notes

1. *You Want a Returnable Bottle? Are You Some Kind of Nut?*, from Environmental Action, Dupont Circle Bldg., Suite 724, Washington, DC 20036.
2. "For the Bottle Bill: This Could Be the Year," *Environmental Action,* February 8, 1975.
3. "Throwaways: Please Dispose of Entirely," by Patricia Taylor, *Environmental Action,* November 10, 1973.

Additional Sources

"Oregon's Bottle Bill: Two Years Later," by Don Waggoner, Oregon Environmental Council, Portland, OR, May 1974.
The Case for Returnables, Environmental Action Foundation, 724 Dupont Circle Bldg., Washington, DC 20036, 1975.

#49 SORT TRASH

Most solid waste has some value, whether it is reused in the home, recycled for industrial purposes, or composted and mulched in the garden. Before this can be accomplished, however, a workable plan for sorting household trash must be developed. A sample plan is:

- Newspapers: Tie in bundles, take them to a recycling center.
- Glass jars and bottles: Wash and remove labels. When dry, keep in the pantry for food storage or recycling.
- Steel: "Tin" cans can be identified by magnetism, flat tops and bottoms, a side seam, and paper labels. Wash, remove labels and ends, flatten, and take to the nearest recycling center.
- Aluminum: Aluminum cans are identified by their lack of magnetism, light weight, rounded bottoms, pull tops, and lack of side seams. Wash, drain, crush, and take them to a recycling center.
- Plastic bags: Wash and reuse for storing non-food items and garbage.
- Blank pages and junk mail: Use for scratch paper, telephone notes, etc.

- Old envelopes: Use for filing and reuse.
- Paper bags: Use for lining trash cans and cabinets, wrapping packages, and keeping foods in the freezer.
- Old magazines: Donate to neighborhood libraries or give to neighbors and friends.
- Cardboard boxes: Good for storage and children's toys.
- Worn linens: Reusable as napkins and pillow sheets.
- Old clothing: Good for dust cloths, cleaning, and mopping. Use nylons as pillow stuffing (see entry #54).
- Leaves, weeds, grass clippings, wood chips, etc.: Make into compost.

In addition to these suggestions, the consumer must also pressure manufacturers and retailers to make and sell longer-lived and recyclable materials. This can be accomplished by refusing to buy those products that are of poor quality or not recyclable. For companies that market recycled products write to: Environmental Educators, Inc., 1621 Connecticut Avenue, NW, Washington, DC 20009.

References

Recycling: An Interdisciplinary Approach to Environmental Education, by Thomas Fegely, Rita Reemer, Lyn Miller Rinehart, Rodale Press, Inc., Emmaus, PA, 1973.

Solid Waste Disposal Projects, Superintendent of Documents, U.S.G.P.O., Washington, DC 20402, 1973.

Additional Sources

"Solid Waste Information Retrieval System," U.S.E.P.A., Office of Solid Waste Management, Rockville, MD 20852.*

Recycle, from the League of Women Voters of the U.S., 1730 M St., NW, Washington, DC 20036.

National Center for Resource Recovery
1211 Connecticut Ave., NW
Washington, DC 20036

#50 FLUSH-LESS TOILETS

Thomas Crapper's nineteenth-century invention, the flush toilet, may soon join the automobile in technological obsolescence. The flush toilet is responsible for almost half of an average household's consumption of water. A family of four uses close to 90 gallons of water a day for flushing toilets. Clean water is thus used to carry organic wastes to treatment plants. The consequences of this disposal can be extremely damaging both to persons and their environment.

- Pure drinking water is wasted.
- Bodies of water are polluted.
- Septic tanks can pollute ground and well water.
- Waste-treatment plants must be built and maintained to handle the increasing flow of excreta. These are expensive, unreliable, energy-intensive, and only partially clean-sewage effluent.
- Energy and nutrients are wasted that could be returned to the soil.
- The sludge created by "purifying" waste water creates a new problem because it contains organic matter, chemicals, heavy metals, and poisons that cannot be properly disposed. Burning it pollutes the air, dumping it pollutes the water, and using it as landfill can pollute groundwater sources.

These economic, energy, and environmental costs are not

necessary since several non-water toilet systems are available in the United States.

The compost toilet is one viable alternative that is entirely self-contained; therefore, human waste cannot enter the soil or foul waterways. By combining heat, oxygen, and humidity, these units provide a favorable environment in which human waste, toilet paper, and in some cases household garbage can decompose naturally. The result is an odor-free humus that can be spread on the home garden.

One compost toilet, the Clivus Multrum, utilizes aerobic decomposition to convert human excrement and kitchen garbage into humus that is free of harmful bacteria and viruses. The humus weight represents 5 per cent of the original waste, while the other 95 per cent turns into carbon dioxide and water vapor that is vented through an exhaust pipe. The ventilation pipe eliminates any odor by sucking air down from the toilet and kitchen intakes and up through the outlet vent.

A Clivus Multrum can be ordered for around $1,000 from Clivus Multrum USA 14a Elliot St., Cambridge, MA 02138. The company expects the price to drop with mass production. Compare this cost with the price of conventional equipment. A septic tank and toilet installation can run over $1,500.[1] Clivus Multrums also eliminate the costs and energy involved in garbage disposal (including land costs for landfill sites), do not create pollution, and require no maintenance, and no energy to operate.

The Clivus Multrum can decompose human excrement, sanitary napkins, disposable diapers (minus the plastic), shredded paper, and food wastes. Kitchen waste lacking sufficient cellulose must be augmented with leaves, straw, or grass cuttings for proper decomposition. The Clivus Multrum is currently used in Sweden and has the enthusiastic support of Swedish health officials. Last year, Maine approved the Clivus Multrum as an alternative to the flush toilet.

The Ecolet is another tried and proven ecological, compost toilet with over 50,000 in use around the world. The Ecolet's decomposition process is accelerated with an assist from modern technology. Instead of using only the heat generated within a pile of waste, additional heat is provided by coils

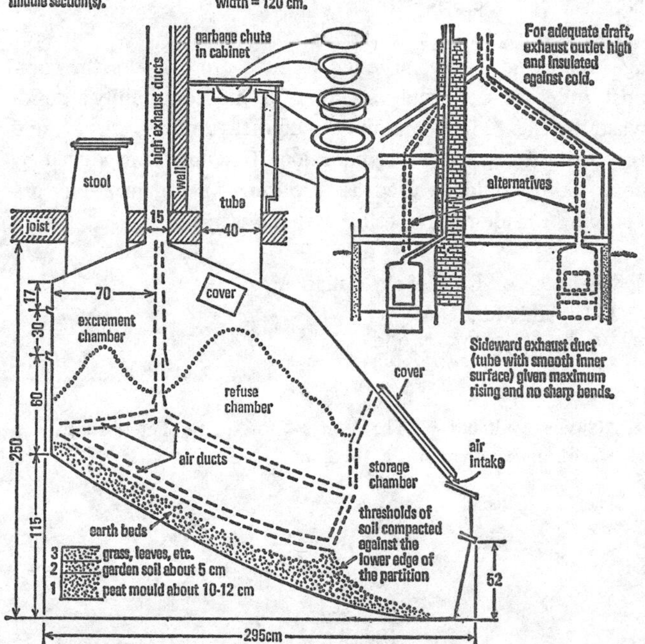

located inside the unit. Oxygen is forced through the pile by an electric fan that draws air from an intake and exhausts it through a wall or ceiling vent, thus eliminating odors.

A starting layer of peat moss contains the soil bacteria required to interact with the bacteria in human waste. Urine adds moisture to the waste pile, and the mixing of human waste and toilet paper allows decomposition to occur naturally and continuously as the Ecolet is used.

The Ecolet costs around $640 (less than ⅓ the price of a septic system). It needs no maintenance or chemicals and is odor-free. The Ecolet uses as much energy per month as three 60-watt bulbs—and even less in warmer climates. For more information contact: Recreation Ecology Conservation of United States, Inc., 9800 West Bluemound Rd., Milwaukee, WI 53226; or Ecolet: Thorton Gore Associates, Box 126, North Woodstock, NH 03262.

If a community is not already committed to a federal sewage treatment project, urge local officials to consider the compost toilet as a possible solution to the community's solid-waste problem. The community could save water, energy, and the costs of organic-waste pollution. The humus produced by composting could be used as fertilizer. An organization concerned with alternative toilet systems is:

Institute for Local Self Reliance
1717 18th St., NW
Washington, DC 20009

Note

1. "Clivus Multrum vs The Water Closet," by Pete Seeger, *Environment,* September 1975, pp. 33–34.

Additional Sources

Other Homes and Garbage, by Jim Peckie, Gill Mastas, Harvy Whitehouse, Lily Young, Sierra Club Books, San Francisco, CA, 1975.

Stop the Five Gallon Flush: A Survey of Alternative Waste Disposal Systems, Minimum Cost Housing Study Group, McGill University, Box 6070, Montreal 101, Can.

Compost Privy, by Sim Vander Ryn, Farallones Institute, Box 700, Point Reyes, CA 94956.

Methane Digesters for Fuel, Gas, Fertilizer, New Alchemy Institute, Box 432, Woods Hole, MA 02543.

Composting in the City, from the Institute for Local Self Reliance, 1717 18th St., NW, Washington, DC 20009.

VI CLOTHING

INTRODUCTION

Clothing can be an expression of personality. A flashy dresser may be gregarious and outgoing, while a casual dresser may wish to remain low-keyed and avoid unnecessary frills. But regardless of personality or expression, it is possible to simplify one's clothing collection without sacrificing style.

This section contains suggestions for: choosing fabrics; buying shoes and quality merchandise; making, mending, and redistributing clothes; and reasons to refrain from wearing furs from endangered species. Following these suggestions can reduce one's demand for clothing and thus reduce resource depletion. These measures can also provide better appearance, financial savings, and personal satisfaction.

To improve clothing choices, it is best to know the facts. For example, clothing materials, whether synthetic or natural, require natural resources–land, energy, and human labor. Growing cotton, for example, is the third largest consumer of energy among U.S. crops (176 trillion Btu's annually). Additional energy is required to transport, process, weave, fabricate, and send materials to commercial markets. The cost in human terms is considerable (brown lung disease among cotton workers is common).

Other fabric and garment expenditures are also resource costly, especially those of synthetic fabrics derived from nonrenewable petrochemicals. From an energy resource standpoint, the least costly is wool, since the sheep are grown on lands that are sometimes incapable of raising crops. Thus, there are ways to conserve energy and resources depending on the type of clothing that is worn.

#51 CHOOSE FABRICS WISELY

Clothing fabrics were once limited to natural fibers such as cotton, wool, flax (linen), and silk. Cotton was inexpensive, but was obtained at high human energy costs to both southern slaves and free men. Wool, one of man's oldest fibers, was too heavy for warmer American climates. Flax never gained the popularity it found in northern Europe, while silk was always expensive.

In the twentieth century, synthetic fibers have often replaced natural ones in clothing and textiles. Cotton dropped from 30.9 lb. per capita in 1950 to 14.18 lb. per capita in 1975; wool from 4.2 lb. to 0.52 lb.; flax from 0.07 lb. to 0.02 lb.; silk from 0.07 lb. to 0.01 lb. Although rayon and acetate, the first commercially available man-made fibers, fell from 8.9 to 3.75 lb. per capita, the non-cellulosic varieties (nylon, polyester, acrylic, and others) rose from 0.9 lbs. per capita in 1950 to 30.10 in 1975.[1]

The division between natural and man-made fibers is becoming blurred as new fabrics appear in blended forms. Cotton and rayon combined with others will improve absorbency, comfort, and dyeing ability while reducing static build-up and production costs. Nylon adds strength to fabrics, as proven by a nylon rope's superiority to one made of hemp. When added to blends, polyester contributes wash-and-wear qualities and wrinkle-resistance.

A consumer looks for certain qualities when purchasing fabrics and clothing: resistance to shrinkage in shirts, durability in socks and play clothes, lightweight in windbreakers, absorbency in underwear, colorfastness in draperies, and fire-resistance in sleepwear. Certain qualities are fashion-inspired, such as sheen and shape in hosiery and crease-resistance in trousers.

Some ecology-minded people follow this general rule—"Use natural fibers." Wool is long wearing and warm; cotton is absorbent and soft; linen is durable and heat-resistant. However, this general rule has major exceptions. Synthetic socks usually last longer than pure cotton ones; trousers made from blended materials tend to hold a crease better; coats made from wool and synthetics have better wind- and water-repellency. Thus, a better rule is—"Use pure or blended materials according to the qualities sought in the item" (see chart).

According to a study by economist L. B. Gatewood, the total energy consumption from raw materials to finished fabric for one pound of cotton fabric is considerably less than an equivalent weight of synthetic fiber.[2] In manufacturing alone, natural fibers are favored because they do not use non-renewable resources (petrochemicals), whereas synthetic (nylon, polyester) production does. However, if synthetics last longer than cotton, the advantage is lost. Additional savings may be gained in washing, ironing, and drying the products. A few hints might include:

- Do not purchase fabrics only for fashion reasons.
- Select lightweight and permanent press for travel.
- Wear wool for warmth.
- Wear cotton for absorbency.
- Wear specialized synthetics for very hard wear (such as socks and children's clothing).

Notes

1. *Cotton and Wool Situation,* U.S.D.A. Economic Research Service, 1976.
2. *The Energy Crisis: Can Cotton Help Meet It,* by L. B. Gatewood, National Cotton Council of America, January 1974.

	Cotton	Linen	Wool	Silk	Acetate	Triacetate	Acrylic modacrylic	Nylon	Olefin	Polyester	Rayon
Absorbency	+	+	+	+	–	–		–		–	+
Appearance & hand				+	+	+	+				+
Colorfastness	+	+	+	+	–		+	+		+	+
Dimensional stability		+		+	+	+	+/–	+		+	–
Drapability				+	+	+					+
Durability		+									
Dyeability	+		+	+					–		+
Elasticity								+			
Laundrability	+										
Pressed-in crease retention	–	–	+	–	–		+			+	
Resiliency			+				+	+		+	–

Resistance:											
abrasion		−		−	−	−	−	+	+	+	−
ageing				−					+		
chemicals			−				+		+		
friction			−								
heat	+	+		−		+					
mildew	−	−			+		+	+		+	−
moths	+	+	−		+		+	+	+	+	
perspiration	+	+	−	−				+	+	+	
pilling			−				−	−	+		
stains									+	−	
sunlight	−	−	+	−			+	−	+	+	
weather									+	−	
wrinkling	−	−	+	+	−	+	+	−	+	+	−
Softness	+						+				−
Strength Wet	+			−	−	−	−	+		+	
Dry	+			+	−	−	−	+		+	
Warmth			+				+			+	
Wash & Wear quality	−		−		−	+	+			+	−
Water Repellency			−								

SOURCE: "Fibers and Fabrics," by J.M. Blandford and L.M. Gurel, Consumer Guide from the National Bureau of Standards: Gaithersburg, MD 20760, 1970.

+ = good to excellent
− = fair to poor

Additional Sources

Fiber: A Bibliography, American Crafts Council, New York, NY, 1975.

Warm as Wool, Cool as Cotton: The Story of Natural Fibers and Fabrics and How to Work With Them, by Caster Houck, Seabury Press, Inc., New York, NY, 1975.

Butterick Fabric Handbook, A Consumer's Guide to Fabrics for Clothes and Home Furnishings, by Irene C. Kleebert, Butterick Publishing Co., New York, NY, 1975.

#52 PICK QUALITY CLOTHING

The clothing industry employed 2.4 million Americans in 1973. Another 800,000 are employed in wholesale and retail sales. The garment industry is important and necessary, but the quality of items made and sold vary considerably. Some clothes never wear out and others hardly last a day. With about $200 per person used for clothes each year, this becomes a major consumer item in the budget.

Quality clothing generally refers to something well made that will last. However, name brands and high prices do not always indicate top quality, and wise shoppers can find good-quality clothing at bargain prices.

When buying clothes, one should first examine the material's strength, checking for weak spots and a sturdy weave. Some poor-quality cotton and wool fabrics use a chemical dressing to stiffen the material, which washes out and leaves a limp garment.

Check the fiber content, as blends often give longer wear, retain shape, and are easily cared for. One should also check the hang of the material, since some fabrics lose shape and wrinkle easily. If a garment doesn't look good on a hanger, it probably won't look good on the buyer.

One should always try on clothing to check the fit. Examine the inside of the garment and be sure that facings (neck, arms, cuffs) are flat and not buckled. There should be allowances for about ½″ on the side seams and 3″ on the hem. This ensures that one can let out a seam or hem should styles or body figure change. Also, it shows that the manufacturer did not skimp on the product. Check the seam stitch to make sure it is straight and not pulling from being too tight.

Zippers, buttons, or snaps should not pull, and when wearing the garment, these should close without straining the material or the person wearing it. Some garments such as pure wool pants, skirts, or jackets should be lined to reduce irritation and provide better fit. Proper fit is important because durability is reduced if seams are pulled or the garment binds.

Loosely woven material and pure wools sometimes need linings to prevent sags at the knees or elbows. Do not confuse bonded fabrics with lined materials. Lined garments usually provide a better fit, letting the garment mold to the body, while bonded materials are usually stiff. A worn-out lining can be replaced if the garment is still in good condition, but worn-out bonding cannot.

A garment's life can be lengthened with proper care. Before purchasing any clothing item, read the label for laundering directions. Hand washing and dry cleaning require additional expense and time. Some clothes cannot be dried in direct sunlight, and others require ironing. Follow the label instructions.

Air drying clothes in the house can save fabric wear and tear while conserving energy and increasing home humidity. Proper storage is also important. Make sure items are thoroughly dry to prevent mildew. Pack wools with moth balls.

We buy different clothing items for different reasons, such as jeans for durability and pants for fashion. "Fad" clothing is a waste of money because it fades quickly out of style. One should purchase basic outfits that can be interchanged with other clothes. Purchasing a few basic outfits that can be used for years will also spread out the original investment. Since conservation also refers to saving money, watch for sales on well-made clothing.

Buying Hints

- Pick garments best suited to needs and required care.
- Check sizes and fit. The right fit avoids rips and tears.
- Study style features and trimmings to be sure they will last in use.

- Examine the garment's workmanship outside and inside, to make sure it is appropriate for the material, style, and use it will get. Always look for flaws.
- Take time to pick the best garments. While one choice may seem as good as another, clothes are made by individuals, some of whom are more skillful and exacting than others.
- If ever dissatisfied with a purchase, return it to the retailer or write the manufacturer.

#53 MAKE CLOTHING

Bolts of fabric and yards of trim are a wonderland for the imaginative person. Creating everything from suits and dresses to slipcovers and garment bags is possible. Sewing and knitting are productive hobbies that offer opportunities to relax and develop a skill. Homemade clothing generally affords a tailor fit, wears longer, and costs less than a factory-made article. In this way, quality clothing with a tailored fit can be acquired without paying exorbitant prices.

Making clothing gifts can be rewarding. With a little ingenuity a designer-type tie can be made for under $3.00 or a scarf and hat set for about $4.00. A creative project might include making a jacket from a bulky plaid blanket, curtains from flowered sheets, or a robe using velour towels. Patchwork quilts, pillows, skirts, and shirts can be easily fashioned.

Doll clothes or improvised bandanas or slings can be made for children's games.

To sew, one needs a work place with good lighting and the following equipment:

- Basic sewing course
- Sewing machine, footpedal models save energy
- Iron, pressboard, press cloths, and sponge
- Scissors–make sure points are sharp
- Pinking shears–use to finish the edges of fabrics
- Ruler–preferably transparent
- Flexible tape measure
- Pins
- Needles–get crewel type (with long eyes)
- Thimble
- Embroidery hoops
- Ripping aids
- Hooks and eyes
- Snap fasteners
- Buttons
- Threads–for heavy and light materials
- Buttonhole twist
- Darning cotton and wool yarns
- Tapes
- Press-on interfacing
- Pencils or chalk

To insure good fit it is important to buy the proper-sized pattern. Buy material to match pattern style. For instance, a splashy print may not look good in a sheath-styled dress but might look great in a full-skirted dress. When choosing a pattern, the weight of the material should be considered. Don't try to use heavy material, such as duck or heavy wool, for a pattern with many seams, gathers, or detail. Often types of materials are suggested by the patternmaker. This service is a good aid for beginners to follow. Note cutting and layout directions. Try on the garment occasionally so that fit can be adjusted. Press the garment as it is being made so that after making darts, joining facings, etc., the work is flat and easy to work with.

There are certain indispensible stitches that ought to be learned. The following stitches are valuable in finishing or mending a garment. Fine handwork adds the finishing touches to the creation.

Hemming stitches: The running stitch (Fig. 1) is especially good if spaced stitches are needed. A whipping or slanted stitch (Fig. 2) works best if close stitches are required. Notice that the thread in the running stitch is under the hem fold, but is on top in the whipping stitch.

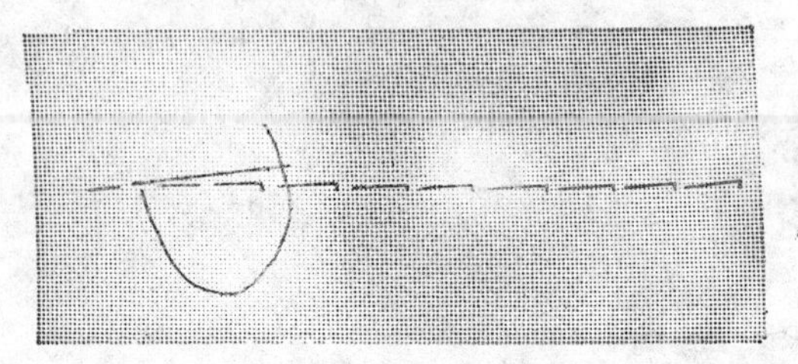

FIGURE 1

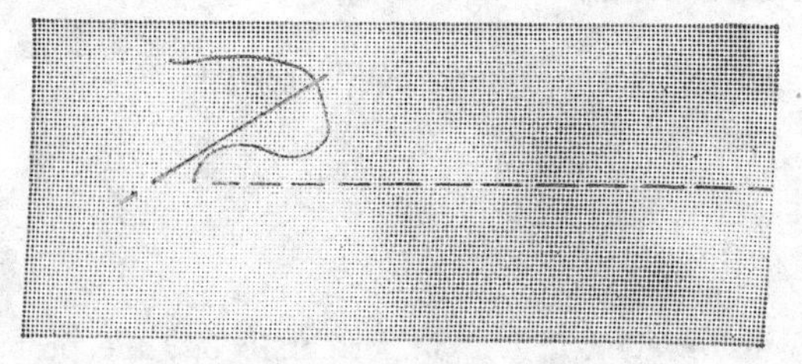

FIGURE 2

Overcasting stitch: Overcasting makes a good seam finish to protect cut edges against ordinary but not excessively frayed material (Fig. 3).

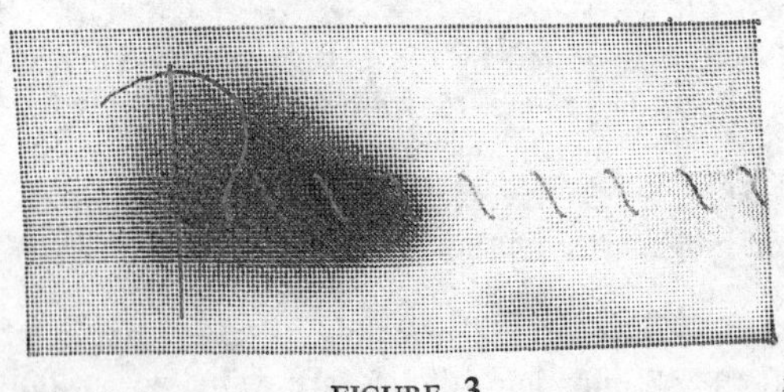

FIGURE 3

Backstitch: For places hard to reach by machine–underarm seams, gussets, and plackets–the back stitch (Fig. 4) gives the appearance of machine stitching. The underneath stitch is twice the length of the top stitch. Top stitching looks like machine stitching because each top stitch meets the next stitch.

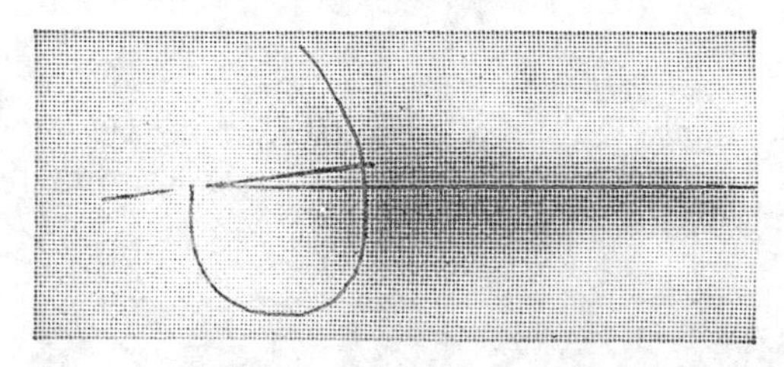

FIGURE 4

Seed stitch: This variation of the backstitch, in which only tiny stitches show on the right side (Fig. 5), is strong but practically invisible. It can be used to repair zippers put in by hand, and in other places where appearance matters. A long underneath stitch permits a space between small top stitches.

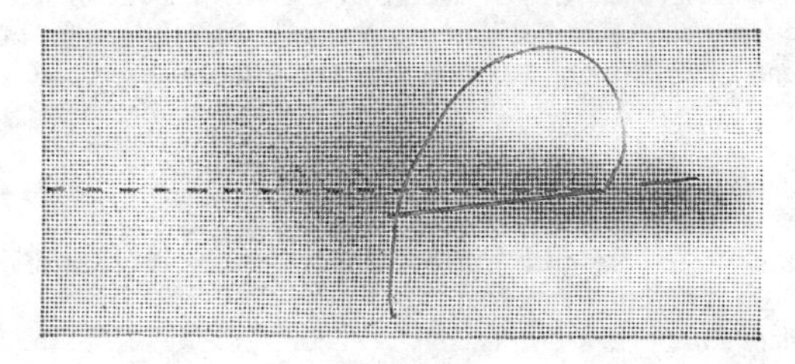

FIGURE 5

Padding stitch: The padding stitch (Fig. 6) is helpful for tacking and holding two layers of fabric in place before machine darning. It also reinforces a darn and protects against inside wear.

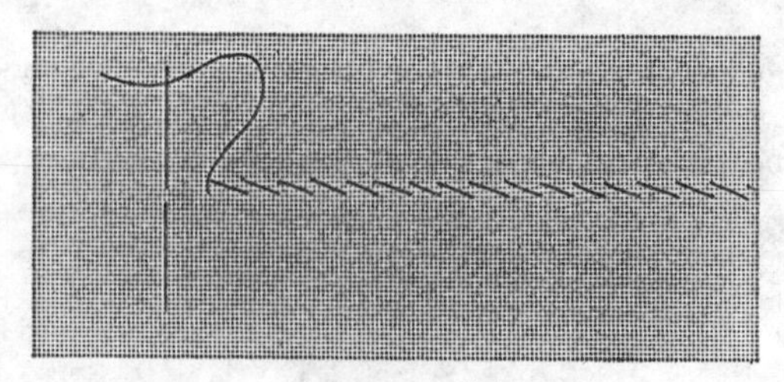

FIGURE 6

Blanket stitch: The size of a blanket stitch depends on its use. Make it large for edge finishing (Fig. 7) and very tiny for strengthening weak corners.

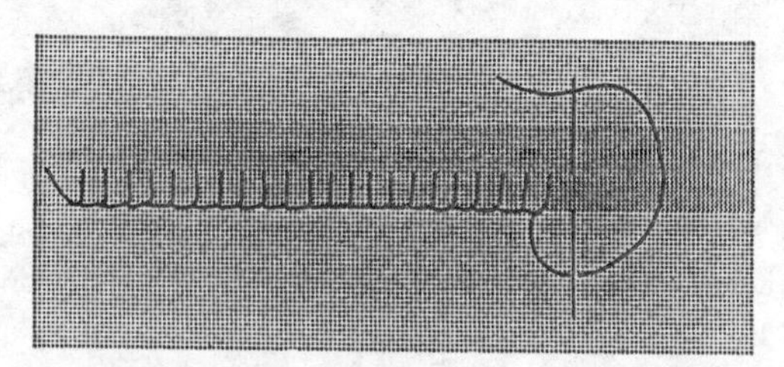

FIGURE 7

Reference

"Clothing Repairs," USDA Home and Garden Bulletin, No. 107, from the *1973 Yearbook of Agriculture,* Superintendent of Documents, U.S.G.P.O., Washington, DC 20402.

Additional Sources

The Penguin Book of Sewing, by Julian Robinson, Penguin Books, New York, NY, 1974.

Hassle Free Sewing and Son of Hassle Free Sewing, Straight Arrow Books, San Francisco, CA.

American Patchwork Quilts, by L. Bacon, Monow, William & Co., West Caldwell, NJ, 1973.

The Beginners Guide to Basic Sewing, by Tad Rowady, New American Library, New York, 1974.

Singer Sewing Book: The Complete Guide to Sewing, by Gladys Cunningham and Jessie Hutton, Random House, Westminster, MD, 1972.

Children's Clothes: Easy to Make Clothes for One-to-Ten-Year-

Olds, by Jill Morris, Arco Publishing Co., New York, NY, 1975.
A Complete Guide to Home Sewing, by Sylvia K. Mager, Pocket Books, New York, NY, 1973.

#54 MEND AND REUSE GARMENTS

Conservation of resources and avoiding waste is inherent in any simple lifestyle. To avoid waste extend the life of clothing by mending, reusing, or redistributing garments. Remember to fix something before damage becomes extensive. Otherwise major repair jobs will be necessary. A few pointers for mending, patching, and darning operations are:

Mending Pointers

- For buttons that have pulled off and ripped the underlying material, sew a small patch on the inside of the garment to reinforce before sewing the button on.
- Zippers can be replaced securely and ripped stitches can be fixed by hand using a backstitch.
- If sock toes and heels are worn beyond the darning stage, sew patches of a stretchable material that breathes.
- If a favorite sweater or nylon shirt has a long run in it, consider hiding the run by embroidering over it.
- Cigarette burns or holes sometimes can be covered with appliques.
- When pants or shorts are ripped at the seams so that no

seam allowance is left, put in inserts of a different material.

Patching Pointers

- Patch faded garments with similarly faded material.
- If new fabric must be used to patch a washed and shrunken garment, shrink the patch piece.
- On a ready-made garment patch material can usually be taken from a facing, hem, pocket, or sash.
- If the fabric has a design, slide patch material beneath the hole until the pattern matches. For material with an up-and-down pile, match the direction of the pile.
- Cut a patch with the grain of the material making sure the grain directions match those in the material being repaired.

Darning Tips

- Study the weave of the original fabric and reproduce it in the darn.
- Work under good lighting.
- Use a fine needle and short thread. A long thread, pulled back and forth across a torn place, may pull the area out of shape.
- Darn on the right side of material and blend the darn into the fabric around the hole.
- Work for flatness. If yarns are pulled tight, the finished darn puckers. Loose stitching gives yarn a puffy look.
- Take small stitches. Do not draw the thread or yarn taut when making a turn. Run the stitching unevenly into the cloth around the darn to prevent a heavy line around the darn.
- Pull the ends of darning yarn to the underside of work and cut off, but not too closely. Be sure all raw edges are on the underside.
- Steam press finished darn on the wrong side. If the material is wool or napped, brush darn to lift up nap.

To Make a Neat Machine Darn

- On a flat surface, straighten and trim tangled or frayed yarns.
- Cut an underlay from lightweight press-on interfacing

fabric, large enough to reinforce and hold the torn area in place.

- Slip the reinforcement underneath the area, adhesive side up. Hold in place with pins. With yarns combed precisely in place, cover mend with a thin cloth to protect fabric, then press.
- With matching thread in a slightly darker color, machine stitch back and forth over the damage with the grain.
- Trim away any excess of the reinforcing material, unless the surrounding area needs it for strength.
- Tack reinforcement invisibly to the back of the fabric with padding stitches.
- If the damage is a three-corner tear, or is badly frayed, machine stitch both crosswise and lengthwise.

There are many ways to reuse clothing. An obvious suggestion is to pass an article down to younger or smaller family members. Another is to alter or repair the garment in some way to make it wearable. A dress can be cut to make a blouse or smock top, pants can be cut for shorts, or an old quilt can be used to make a new purse. If an item is unusable, don't throw it away, cut it up for cleaning rags. Turkish toweling makes good scrub rags, cotton items are excellent for polishing, and wool works fine as a dust cloth. Remember clothes may be donated to charity groups such as: Goodwill, Salvation Army, St. Vincent de Paul, and various thrift shops. Support neighborhood clothing drives and thrift shops. Theater groups also are eager for clothing donations.

Reference

"Clothing Repairs," USDA Home and Garden Bulletin No. 107, from the *1973 Yearbook of Agriculture,* Superintendent of Documents, U.S.G.P.O., Washington, DC 20402.

Additional Sources

Living Poor With Style, by Ernest Callenback, Bantam Books, New York, NY, 1972.

Gladrags; Redesigning, Remaking, Refitting, by Delia Brock and Lorraine Bodger, Simon and Schuster, New York, NY, 1974.

The Beginner's Book of Patchwork Quilting, by Lydia P. Encinas, Bantam Books, Des Plaines, IL, 1974.

A Book of Great New Ideas for Decorating Jeans and Jackets, by Eve Harlow, Drake Publishers, Inc., New York, NY, 1973.

Caring for Your Dress, by M. Middlemas, Pergamon Press, Inc., New York, NY, 1967.
Make Your Own Alterations: Simple Sewing the Professional Way, by Miriam Morgan, Arco Publishing Co., New York, NY, 1970.
The Personal Touch, Time-Life Books, New York, NY, 1974.

#55 DO NOT WEAR FUR FROM ENDANGERED SPECIES

Conserving natural resources in the clothing industry can be aided by consumers if they do not choose clothes made from the fur and hides of endangered animals. Each year, more than 25 million wild animals are trapped as victims of a fur trade that caters to fashion. In the United States and Canada alone, 13 million animals including beavers, lynx, otters, seals, squirrels, wolves, foxes, mink, muskrat, and raccoons are killed for their fur.[1] On the international scene the world trade in spotted cat skins is endangering their very existence.

The steel-jaw trap is the primary means by which North American trappers capture their prey. Once caught, animals may struggle for weeks to free themselves, sometimes gnawing off limbs to escape. Many non-fur-bearing animals also fall victim to the steel traps. Trappers commonly refer to these animals as "trash." A five-year Canadian survey of two traplines showed that of 1,911 animals trapped, only 561 had usable fur; other animals, including geese, ducks, song birds, eagles, owls, and porcupines were crushed to death in the traps.

Spotted cats, including the tiger, leopard, jaguar, cheetah, and ocelot, are threatened with extinction because of the demand for fashionable fur coats. A majestic animal like the leopard, one of the world's most intelligent animals, and the cheetah, the fastest land animal in the world, could be lost forever simply because wearing their fur is stylish.

One way to have natural furs without the horrors of the steel-jaw trap or the danger of permanent extinction is to use fur from farms raising mink or ermine. Another way is to wear fake furs made from modified acrylic fibers. They closely resemble animal fur, weigh half as much, and wear equally as long. Quality and price vary greatly in the mock-fur market. But the consumer has the advantage because virtually all synthetic furs are made from the same kind of fiber. The difference between brands is strictly in construction and styling. These differences can be seen. Careful examination will tell one about the quality and, therefore, the price of a fake fur. Still another answer is a natural fiber like wool which can be as warm, stylish, and far less expensive than fur, especially in terms of natural resources.

Note

1. "Let Us Live," pamphlet from Defenders of Wildlife: 2000 N St., NW, Washington, DC 20036.

Other References

Fibers and Fabrics, A Consumers' Guide from the National Bureau of Standards, U. S. Commerce Department, Superintendent of Documents, U.S.G.P.O., Washington, DC 20402.

No Skin Off Their Backs, from The Wildlife Fund: 1319 18th St., NW, Washington, DC 20036.

"The Next Best Thing: The Truth About Fake Furs," *Consumer Gazette,* December–January 1975.

"Fur Issue," *Defenders of Wildlife Magazine,* October 1975.

Extinct and Vanishing Animals, by Vizenz Ziswiler, Springler-Verlag, New York, NY, 1967.

Additional Sources

Wildlife in Danger, by James Fish, Noel Simon and Jack Vincent, Viking Press, New York, NY, 1969.

Wildlife Management Institute
709 Wire Bldg.
1000 Vermont Ave., NW
Washington, DC 20005

National Wildlife Federation
1412 16th St., NW
Washington, DC 20036

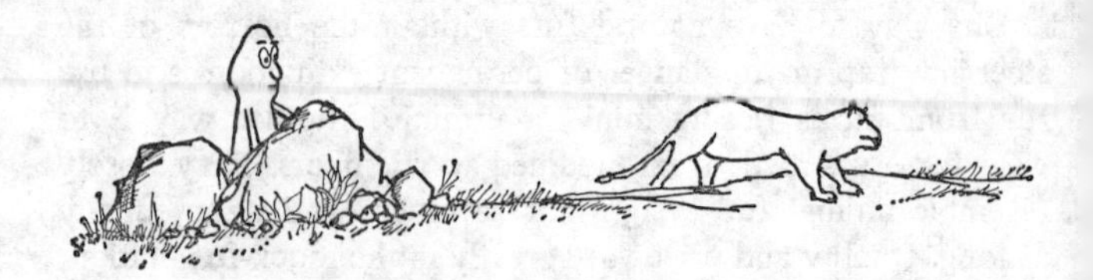

#56 PROTECT THE FEET

When buying shoes, three considerations are: 1) Do they fit? 2) Is price consistent with quality? 3) Is style compatible with healthy feet? Automation in the shoe industry has made shoes less expensive and easier to obtain; but finding shoes that fit properly is always a problem. Shoe salesmen are not usually professionally trained, so the responsibility rests with the buyer. When buying shoes for adults, the following rules generally apply:

- Measure the foot while standing. Use a brannock device that measures the foot from heel to ball.
- With the foot in the shoe, place a finger inside the shoe at the ball and try sliding the finger. If there is room the shoe is too wide.
- Stand and raise the heel off the floor. If creases run on an angle toward the big toe joint, the fit is improper.
- Return to seller if the feet hurt after wearing the shoes.

Fitting shoes for children is often difficult because their feet grow rapidly. It is important to get the proper fit, since a child's foot is mostly cartilage and is very malleable. Therefore, the foot will conform to the shape of the shoe. Children's shoes are fitted individually—do not pass them on to be worn by another child.

It is very important that children's shoes not be too tight. There should be at least one-half inch or more growing room in the toe. Check thc child's shoes regularly.

To judge price with quality, one should know the characteristics of well-made shoes:

- Close conformity to the foot's natural shape.
- Neat finishing with no rough edges, loose threads, or loose heels.
- No paper-thin soles or exposed nails or tacks.
- Smooth inside surfaces.
- Reinforcements at stress points.

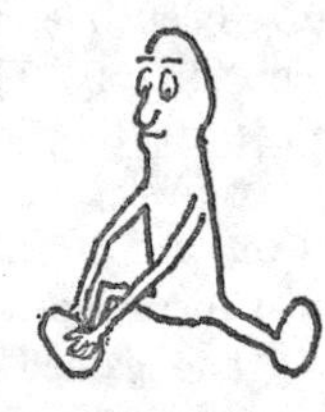

- Welt construction (a strip of leather between the upper portion and the sole that allows for easier repairs).
- Genuine leather sock lining.
- Leather overlaps at the top of the heel seam prevent splitting.

Leather is generally the most desirable material for shoes. Good-quality leather will be pliant, firm, shape retaining, and durable. It is also porous, allowing for ventilation. Calfskin is probably the most durable. The appearance of little lines when the shoe is flexed is a good sign.

Plastic generally is not a satisfactory material for shoes since it lacks porosity and creates hot, damp foot conditions. If used, plastic should be packed with fabric. Two improved plastics, Corfam and Aztran, are porous, water-repellent, lightweight, come in various colors, require no polishing, are scuff resistant, and mold to the feet. However, they are usually more expensive than leather.

Shoe soles should be made of thick leather or hardened rubber. Rubber heels are more durable and resilient than leather. For sports, dress, or home wear, shoes can be made using materials ranging from nylon to cotton, and using fabrics from gabardine to brocade. Many of these materials are elasticized to provide greater comfort and shape retention.

One should not buy extreme shoe styles such as very high heels or platforms. Platform shoes with thick heels can cause walking and driving accidents. Physicians report that people who wear platform shoes can suffer from sprained ankles, pelvic distortions, disc problems, and fractured feet. It is difficult to keep one's balance when elevated, and high heels can contract the calf muscles so that wearing low heels is uncomfortable. For the best fit, a shoe should conform to the foot's natural shape.

Before buying a negative-heel shoe, one should be sure that the shoes can be worn comfortably. The heel on this shoe is lower than the sole, the toe area is wide, and the arch is elevated. Most users need a two-week adjustment period to ease and reduce heel pain as the calf muscles become shorter. A sample of people who wear these shoes found that 70 per cent walked more comfortably and gained relief from corns and bunions. Podiatrists believe these shoes can help people with moderately flat feet and those recovering from foot surgery. However, people with very flat feet, high arches, shortened calf muscles, or short Achilles tendons are advised not to use negative-heel shoes.

Tennis shoes are the best for comfort. They are light and cool, and some types are washable. Construction should include foam- or sponge-cushioned inner soles, non-slip treads, and reinforcements at the heels, eyelets, sides, and toes.

Prevent foot problems by always practicing the following:

- Buy shoes at the end of the day when feet are larger.
- Fit shoes properly. Extreme styles can cause friction and pressure on parts of the foot. Calluses, corns, and bunions can result.
- Air feet and shoes daily.
- Do not walk on corns and calluses; see a podiatrist.
- Bathe feet every day. Towel gently and powder between toes.

- Clip toenails regularly; do not use scissors or files on the cuticles. Smooth nails with an emery board.

References

A Really Big Shoe, Consumer Survival Kit, Maryland Center for Public Broadcasting, Owings Mills, MD 21117.

First Clinical Study of Earth Shoe, completed by the California Podiatric Medical Center, Ruder and Finn of California, 100 Bush St., San Francisco, CA 94104.

"Balancing the Load on Your Feet," *Medical World News,* September 20, 1974, p. 56-F.

"The Negative-Heel Shoe—Pro and Con," by Nadine Brozan, the New York *Times,* March 3, 1975.

Additional Sources

Shoes for Free People, by David Runk, University Press, Santa Cruz, CA, 1975.

All About Distance Running Shoes, by the editors of *Runners World,* World Publications, Mountain View, CA, 1972.

VII FULFILLMENT

INTRODUCTION

The ability to create works of art is an innate human talent. With knowledge of form, technique, and material, we can combine these elements in imaginative ways. Through trial, error, and discipline, we can achieve the simple but profound satisfaction of creating a gift for a friend or decoration for the home.

In modern society, there is little opportunity for one to earn a living through handcrafts. Craftsmen have been replaced by machines and assembly lines. The market place, where crafts were once bought and sold–and people mingled with neighbors to exchange news and cultivate friendships–has been replaced by impersonalized shopping centers.

Instead of the eat-and-run events that characterize modern lifestyles, families should spend time together. For a start, a family could gather to share one leisurely meal a day. The television should be turned off to allow family members to tell of their day's experiences and exchange ideas. Once the habit sets in, discussion becomes natural, and dinner-table conversation becomes a major event.

The more time allowed for family gatherings, the better. Sporting events, dancing, painting, sculpture, and other activities can be shared between parents and children as participants or spectators. A child will take greater pride in a pursuit when parents are present. Too much technological gimmickry has obstructed natural communication and the development of creative expression. All too often, a child's creativity is crushed by an attitude that art is for the professionals.

Communication and creative expression should not be limited to the family. The local community is another area where one can promote the positive attitudes of social and intellectual exchange. Citizens must become concerned with community issues. Gather with friends, neighbors, and family to watch or organize community sporting events. Get to know the community environment. If interested in a particular craft or hobby, teach it to others in the community or learn one from a neighbor. Perhaps someone has a musical talent. Share enjoyable books and discuss them with others. The local library is a good place to browse for interesting books, and the library bulletin board is an excellent way to bring attention to a particular organization or individual interest worth sharing with others.

#57 HAVE A HOBBY

The American past is rich in cultural currents, and each contributes to the crafts of different ethnic and racial groups. Many of these crafts are frequently displayed in local and regional festivals. By visiting these festivals, we might find inspiration to start our own hobbies.

We all have a creative urge of some sort. It is a challenge to find new approaches to handcrafts that stimulate our imaginations, curiosity, and latent talents. Practicing a craft can open channels for social contact, and emotional and aesthetic growth. For the beginner, it is best to work with uncomplicated, traditional crafts. The variety is limitless, and there is something for each of us. Here are some simple suggestions:

SALT CERAMIC: For modeling or sculpturing:

1 cup table salt, ½ cup cornstarch, ¾ cup cold water. Mix together in a double boiler and place over heat. Stir constantly for 2 to 3 minutes and remove when the mixture is thick. After it has cooled, knead it for several minutes and it is ready to use. It is an excellent substitute for clay because it does not shrink when drying, will harden to the consistency of stone and does not powder like clay.

OLD INNER TUBES: For printing:

(These function as block prints.) The materials can be found in a junk pile or an abandoned lot. Cut the inner tube into 5-square-inch pieces. Sketch an idea with chalk on the inner tube and then cut around the edges with scissors. Glue the design to the flat side of a block of wood and use like a block print.

PUPPETS:

A variety of items: bags, old socks, needles, yarn, old newspaper for stuffing, colored paper, paints, and a little imagination can permit the expression of a child's wildest dreams. Adults also find it engrossing. The child may paste, staple, or glue the various size bags using them for arms, legs, heads, eyes, ears, noses, or other facial features. One can always bring the puppets to life by staging puppet dramas or plays.

Other hobbies include:

- Modeling and sculpturing using materials such as clay, wire, toothpicks, wax, wood, discarded plastic, and aluminum.

- Print making, using paints, crayons, printing pads, sticks, and paper stencils.
- Painting with water colors on window glass or wet paper.
- Making collages, banners, flags, murals, ceramics.
- Decorating eggs for holidays and special events.
- Additional Ideas: papier-mâché and mask making; weaving, hooking, embroidering, batiking, and appliquéing; general craft building, such as building a stage with boxes, making toys, dolls, models, and kites.

Frequently, the best places to look for materials are in one's community, on the streets, in alleys, junkyards, railroad tracks, playgrounds, attics, or closets. The materials are usually free. It is surprising how many hobbies have been unearthed from collected debris. Keep materials on hand that might be used for a future project. For more information, browse through the numerous books and pamphlets on crafts in a library, or go to a local fair or festival.

References

Meaning in Crafts, by Edward L. Mattil, Prentice-Hall, Englewood Cliffs, NJ, 1965. (Excellent detailed descriptions of many crafts for children.)

Nature Crafts, by Anne Orth Epple, Chilton Book Co., Radnor, PA, 1974. (Decorations made from bark, cones, ferns, feathers, nuts, moss, shells, seeds, etc.)

Additional Sources

Encyclopedia of Needle Work, by Therese de Dillmont, Joan Toggit, Ltd., New York, NY.

A Handbook of Crafts, by Elsie V. Hanauer, A. S. Barnes & Co., Cranberry, NJ.

The Crewel Needlepoint World, by Barbara H. Donnelly, Morgan Press, Milwaukee, WI, 1972.

Creating Rugs and Wall Hangings, by Shirley Marein, Viking Press, New York, NY, 1975.

The Updated Last Whole Earth Catalog, edited by Stewart Brand, Random House, New York, NY, 1974.

The Artist's Handbook of Material and Techniques, by Ralph Mayer, Viking Press, New York, NY.

The Camera, Camera Buyers Guide, Time-Life Books, Chicago, IL.

Easy Magic, by Roy Holmes, Harper & Row, New York, NY, 1974.

American Indian Craft Inspirations, by Janet and Alex D'Amato, J. B. Lippincott, Philadelphia, PA.
Hobbycraft Around the World, by Willard and Elma Waltner, Lantern Press, New York, NY, 1966.
Meaning in Crafts, by Edward L. Mattel, Doubleday & Co., Garden City, NY, 1972. (For children.)
Paper as Art and Craft, by Jay and Ree Newman, Crown Press, New York, NY, 1973.
Weaving Without a Loom, by Sarrita Rainey, Davis Publishing Co., Worcester, MA, 1966.

#58 BECOME ARTISTIC

Daily activities offer an opportunity for self-expression. Baking a cake, humming a song, and walking to work can provide enjoyment.

The arts, too, offer other ways of implementing simple living. Creative expression is possible for everyone and in every activity and can remove the monotony that often pervades daily routines. Art is not limited to professionals, and artistic activity does not have to result in a salable product; it can and should be produced for its own sake.

In this age of increasing specialization, we often suppress basic human impulses. In primitive societies, creative impulses are expressed in songs, dances, the beautification of dwelling places, in designs on weapons and pottery. We should give attention to: dancing and ballet, choral arts and singing, playing a musical instrument, sculpturing and carving, painting and etching, photography, dramatic arts.

Dancing is a simple form of fulfillment. It takes rhythm, a good step, and a desire to mingle with others. Primitive cultures developed exotic dancing forms, meant to express this fundamental joy of sharing experiences. Advanced cultures also have this need, as expressed in ballet, modern and folk dancing.

Learn to play a musical instrument. It need not be something elaborate such as a piano or accordion. A kazoo or a homemade set of drums can be expressive and amusing instruments. Join others who play instruments and start a musical group.

Choral arts have a history as old as musical forms. Barbershop quartets, choruses, glee clubs, and holiday sing-alongs all are part of a simple lifestyle. Fill the home with song. It elevates the spirits and is an economical form of entertainment.

Learning to paint or sculpture may require some instruction, as do many arts. What comes naturally in song may require assistance in the visual arts. Photography requires an initial investment of a camera and film, and perhaps later, a dark room and developing equipment.

The dramatic arts are perhaps more dependent on community and other resources than are most other artistic activities. A stage for presenting a play, a cast, a director, lighting, and stage equipment don't just fall into place. However, they need not be elaborate, and the rewards are usually worth the trouble.

RESOURCES NEEDED FOR ARTS

	Initial Equipment	*Building and Materials*	*Fossil Fuel Energy*
Dancing	nil	nil	nil
Music			
Electric Organ	high	high	high
Piano	moderate	moderate	nil

	Initial Equipment	*Building and Materials*	*Fossil Fuel Energy*
Drums	moderate	moderate	nil
Reeds	low	low	nil
Choral Arts	nil	nil	nil
Sculpturing	low	low	nil
Carving	low	low	nil
Pottery	low	low	nil
Metalwork	moderate	moderate	moderate
Painting	low	low	nil
Etching	low	low	nil
Photography	high	high	moderate
Dramatic Arts	high	high	moderate

Additional Sources

Step by Step Arts and Crafts at Home, by Eve Jordan, House of Collectibles, New York, NY. (For children.)

McCall's Giant Golden Make-It Book, Golden Press, New York, NY, 1953.

Creating With Wood, by Seidelman and Mintonye, Crowell Collier Press, London, 1969.

Creative Drawing, by Ernst Rottger and Dieter Klanter, Van Nostrand-Reinhold Co., New York, NY, 1963.

#59 DECORATE THE HOME SIMPLY

Home decorations are an expression of one's lifestyle. House plants and simple homemade creations utilizing wild flowers, twigs, seashells, rocks, and discarded trash save money, resources, and energy over luxurious decoration.

DECORATIONS MADE FROM SIMPLE MATERIALS

- Arrangements of Dried Flowers and Grasses[1]

  Materials: Mid-summer is the ideal time. Use flowers, grasses, leaves, and pods. Look in the home garden or in the wild. Pick when fresh, buds open but not in full bloom. Colors that preserve best: yellow, orange, pink, and blue.

  Drying methods: Silica gel does a good job and can be reused. Bury the flowers in it for 1 week.

  Air drying: Pick the flowers free of moisture. Strip the leaves. Tie loosely in bunches of 4 or 5 and hang upside down in a dry, warm room. This is a good method for everlasting, strawflower, yarrow, baby's-breath, cockscomb, globe amaranth, goldenrod, Queen Anne's lace, and bells of Ireland.

  Borax drying: Mix 1 part of Borax with 6 parts of white corn meal. The mixture must be absolutely dry. Place the flowers upside down in a paper bag on 2 inches of the mixture. Pour more around them to cover well, but do not have the flowers touching. Hang the bag in a dark, dry room for 1 to 3 weeks. Remove the flowers when dry.

  Storage: After drying, store in a dark, dry room until ready to organize the arrangement.

- Stained Glass[2]

  Take large pieces of cathedral glass and crack into small pieces. Glue onto large sheets of plain glass cut into circles or rectangles, in the shape of a flower. Press out any air. Glue a loop hanger between glass at the top. Fill cracks with grout mixed with black poster paint, using fingers. Smooth a border of grout around the edges. Cover with a clear finish.

- Papier-mâché Easter Eggs

  Press used aluminum foil into an egg shape.
  Cover with papier-mâché and dry in the sun.

Decorate or paint a solid color. Coat with a clear finish.
Arrange in a bed of moss or leaves.

- Egg Tree

  Anchor a tree branch in a decorated can.
  Dye empty egg shells.
  Attach a string or wire hook to the shell and tie onto the branch.
  Top each egg with a colorful bow.

- Woods Treasure Basket

  Gloss an acorn cap and put floral clay in the bottom.
  Fill with small seeds, dried leaves, twigs, vines, or flowers.
  Glue acorns together artistically. Glue basket to a small piece of bark.
  The same thing can be done with a small bottle or a papier-mâché vase.

- Dried Forest Floor

  Glue lichens, roots, small mushrooms, and dried flowers to a piece of wood or bark.

- Dried-flower Plaques

  Take an unfinished plaque of wood and stain the outer rim with wood finish.
  Paint the middle with 2 coats of black paint.
  Make an arrangement of dried flowers, stems, grasses, and leaves glued to the center of the plaque; press with fingers and wipe off excess glue.

- Miniature Wreaths

  Glue an ornamental loop to a 3-inch circle of cardboard.
  Glue acorns alone or acorns with berries, seeds, and leaves to the circle.
  Coat with varnish or clear plastic.
  Add a tiny red bow at the top.
  Can be made in a tree or bell shape.

Notes

1. "Decorations of Dried Flowers," by Tom Stevenson, the Washington *Post,* July 20, 1975, p. E–6.
2. "Making Gifts at Home," by Mary Norman Delaughter in *The Alternative Catalogue,* 2nd edition, from Alternatives, 1924 E. Third St., Bloomington, IN 47401.

Additional Sources

Dried Flower Designs, published by the Brooklyn Botanical Garden, 1000 Washington Ave., Brooklyn, NY 11225.

Alternative Celebrations Catalogue, 3rd ed., Alternatives, 1924 E. Third St., Bloomington, IN 47401.

House Plants for the Purple Thumb, by Maggie Bayclis, Charles Scribner's Sons, New York, NY, 1973.

Greenhouse Handbook for the Amateur, from the Brooklyn Botanical Garden, 1000 Washington Ave., Brooklyn, NY 11225.

Plant Propagation in Pictures, by Montague Free, American Garden Guilds, New York, NY, 1957.

Making Things, The Handbook of Creative Discovery, by Ann Wiseman, Little, Brown and Co., Boston, MA. (For children.)

Creating With Paper, by Pauline Johnson, University of Washington Press, Seattle, WA, 1958.

#60 CREATE TOYS AND GAMES

Doing is a child's way of experimenting with his world. It is not important to produce a thing in the adult sense. A little one needs the opportunity to use his own unique drive to master and to create.[1]

Simple things aid creativity–let the child dream, imagine, and pretend. Early in a child's life, the parent can provide materials and assist the child's becoming involved. Later, a child can select materials and put them to use in his/her own way.

In a child's early life a parent can help stimulate the ability for amusement. Under three months, a baby can be occupied with objects that encourage using his/her senses.

Listening: a ticking clock
a radio playing soft music
Looking: bright colors via posters or cloth tacked on inside of the crib

As a baby is able to grasp objects, make a cradle gym by tying a heavy piece of elastic across the crib. On short lengths of elastic, tie on a large bell, and a rattle; tie toys to a bounce chair or stroller, using short strings or shoelaces. Materials to interest a child include: a metal cup, pot lids, colorfast ribbons, large wooden spoons, ladles.

A baby starting to crawl begins to choose his/her amusement. A sense of self and independence can be nurtured by familiar items rather than by store-bought toys that leave little to the imagination.

Make a small sandpile by putting cornmeal, oatmeal, or uncooked beans into a big metal baking pan.

Reserve a kitchen drawer for old but safe pots, pans, and lids. Make play dough, flour paste, or paint to give the child a suitable medium for play.

- Play dough: 1 cup salt
  1½ cup flour
  ½ cup water
  2 teaspoons oil
  few drops of food coloring (optional)

  Add more flour if the dough is sticky.
- Homemade paints: flour, water, a little salt, and food coloring.

Three-, four-, and five-year-olds are verbal, sociable, and insatiably curious. Their capacity to become involved with the world around them increases remarkably at these ages. Toy possibilities include:

- Wooden blocks: Saw 2″ × 4″ lumber into lengths of 3, 6, 12, and 24 inches. Sand smooth to eliminate splinters.
- Cardboard boxes: tape lids closed.
- Empty cans with dull edges.

In contrast to toys, television is a passive way for a child to learn about the surrounding world. Reading analysts fear it may be spawning a generation of non-readers hooked on instant gratification. Nothing is left for the ears, eyes, or imagination.

Games and make believe are excellent ways for children to dream, imagine, or pretend. Skills and principles of acceptable social behavior come into play and add value to this form of amusement. A few suggestions:

- A simple house–a blanket draped over a chair or card table.
- Bowling alley–save ten ½-gallon milk cartons to use as pins. A small ball or unopened soup can will make a great bowling ball.
- Sock-baseball–use a miniature bat, and a sock within a sock (knotted) as a ball. A backyard, carport, or porch are places to play.
- Puzzles–choose one large colored magazine illustration and paste it on cardboard. Cut into pieces. Make it easy or difficult depending on the child's age.

Invent games with:

- Rope jumping
- Dice: *Basketball*–2 dice (2 colors). Keep score by tabulating eight to ten rolls per half.

Baseball

6–6 is home run
6–5 is out
3–3 is triple
5–2 is walk
Create other combinations for hits, runs, and errors.

Hangman/Word Guessing

Draw a gallows.
Think of a word and draw dashes to represent its letters (– – – – – –).

Hangman/Word Guessing

With each guess of an unusable letter, draw part of a person. If a figure is completed before the word is guessed, he is hung!

History Game

Stand two feet from a box lid divided like this. Toss a button to make historical year (1776). Each player tosses until he/she has a year for which he can make an accurate historical statement. First player to get 3 such years wins the game.

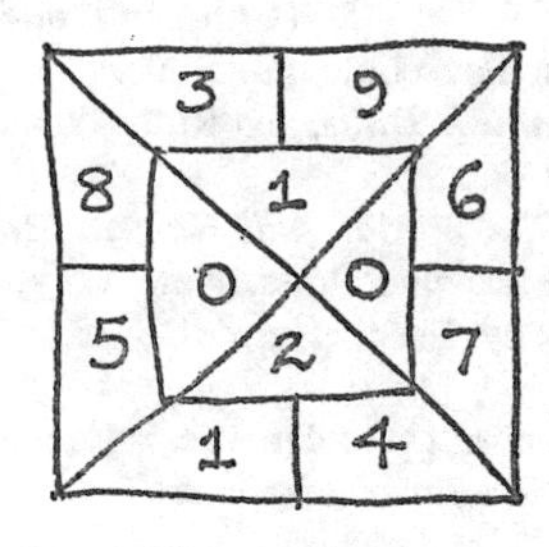

History Game

The Fishing Game

Make a fishing pole from any light stick, string, and bent pin. Fold the fish from paper about 5″ × 7″. Put the fish in a box on the floor. The fisherman stands on a stepladder. Color the fish or assign them values to alter the game. Players take turns. The one with the most fish wins the game.

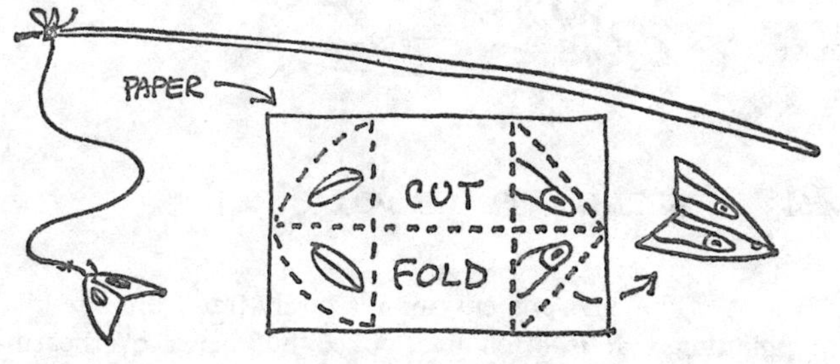

The Fishing Game

Note

1. *What to Do When "There's Nothing to Do,"* by members of the staff of the Boston Children's Medical Center and Elizabeth Gregg, Delacorte Press, New York, NY, 1967.
   An excellent source for elaboration on these and many other ideas.

Additional Sources

The Art of Making Wooden Toys, by Peter Stevenson, Chilton Book Co., Philadelphia, PA, 1971.
Steven Caney's Toybook, by Steven Caney, Workman Publishing Co., New York, NY, 1975. (Children's do-it-yourself projects based on low-cost household materials.)
How to Make Cornhusk Dolls, by Ruth Wendorff, ARCO, New York, NY, 1973.
Folk Toys Around the World: And How to Make Them, by Joan Joseph, Parents Magazine Press, New York, NY, 1972.
25 Kites That Fly, by Leslie Hunt, Dover Publishers Inc., New York, NY 1971.
Kites, by Wyatt Brummitt, Golden Press, Racine, WI 53404, 1971.
How to Make and Fly Paper Airplanes, Bantam Books, Inc., New York, NY.
Children's Games in Street and Playground, Oxford University Press, Fairlawn, NJ 07140.
American Folk Toys and How to Make Them, by Dick Schnacke, Penguin Books, Baltimore, MD, 1973.

#61 PRESERVE A PLACE FOR QUIET

In the urban environment, man is rarely free from one kind of pollution that is often overlooked but never overheard—noise. The blare of city traffic during rush hour, the roar of jets at an airport, the beating of air hammers at a construction site, and the shrill sirens of emergency vehicles all combine to wreck our moments of quiet.

Noise interferes with many daily activities: rest, work, communication, recreation, and sleep. It degrades our quality of life and can damage our health. Depending on its frequency or pitch, noise can impair hearing, raise blood pressure, cause tension and nervous strain, and even alter the rhythm of brain waves so that our thoughts become confused. Loud noises can damage the cilia in the inner ear, making one incapable of hearing sounds and tones vital for daily communication. Constant noises from which we are unable to escape can threaten our sanity.

It is ironic that household appliances, intended to make life more comfortable, are a major source of home noise pollution. Dishwashers, food mixers, vacuum cleaners, dryers, electric shavers, and other appliances are responsible for some of our discomfort. Electric guitars, stereos, radios, and television can also be added to the list of pollutants.

Those most directly affected by high noise levels are industrial workers. The National Institute for Occupational Safety and Health "reluctantly" concurs with the generally accepted 90 dBA (dBA, or decibel absolute, is a unit to measure sound levels: a jackhammer gives 115 dBA, a farm tractor 98 dBA, a 5-ton truck 73 dBA, and a furniture sander 97 dBA). This criteria is meant to "define maximum permissible levels of noise over a period of time which would result in an acceptably small effect on hearing levels over a lifetime of exposure."[1] Yet hearing impairment and other nerve-wracking effects remain as hazards for workers in industry.

The best approach to this problem is to begin in the home. A few suggestions include:

- Use noise absorbing materials on floors.
- Hang heavy drapes over windows to help block out street noise.
- Put rubber treads on uncarpeted stairs.
- Use upholstered rather than hard-surfaced furniture.
- Install sound-absorbent ceiling in the kitchen; wooden cabinets absorb noise while metal ones reflect it.
- Use foam-rubber pads under blenders and mixers.

- Install washing machines in the same room with heating and cooling equipment.
- Use a hand-operated lawnmower; if using a power one, operate it at a reasonable hour.
- Wear a headset when only one person is listening to the hi-fi; keep the volume down with or without the headset.
- Don't buy noisy toys for the children.
- Plant trees and shrubs with dense foliage around the house. Evergreens in particular cut noise from traffic.

To influence change on a greater scale, there are ways to act to curb noise pollution. There are people in the neighborhood who will join in reducing noise levels. Some methods to deal with noise pollution are:

- Pressure local officials to enact ordinances against noise.
- Work toward cutting noise in factories in the locality.
- Force local taverns and nightclubs to curb music after certain hours.
- Stop snowmobiles from entering residential areas.
- Ensure that requirements are met for mufflers on autos and motorbikes.
- Have local school boards direct teachers to inform students of hearing impairment risks from noisemakers.
- Press for quieter mass transit systems.
- Ask police to keep the neighborhood quiet.
- Push for a quiet-time curfew in the locality.
- Redirect approach patterns of noisy airplanes from flying over residential neighborhoods.

Note

1. "Occupational Exposure to Noise," from the National Institute for Occupational Safety and Health, Washington, DC, 1972.

Additional Sources

"Noise—the Third Pollution," by Theodore Berland, Public Affairs Committee, Inc., Washington, DC.
"The Noise Around Us," Report of the Panel on Noise Abatement, U. S. Department of Commerce, Superintendent of Documents, U.S.G.P.O., Washington, DC 20402.
"Noise: The Harmful Intruder in the Home," U. S. E.P.A. Report, Superintendent of Documents, U.S.G.P.O., Washington, DC 20402.
Out of Solitude, by Henri J. Nouwen, Doubleday & Co., Garden City, NY, 1974.
Contemplation in a World of Action, by Thomas Merton, Doubleday & Co., Garden City, NY, 1973.
Meditation: How to Do It, by Alan Watts, Pyramid, New York, NY, 1976.
"Green Revolution," from the School of Living: Freeland, MD 21053.

#62 CAMP AND BACKPACK

Camping is great recreation and is an energy saver, as long as campers do not tote along every home comfort. Simple backpacking is perhaps the most satisfying and enjoyable way to camp. A hiking trip along meandering mountain trails, across the surging seashore, or amid the stark beauty of the desert has a cleansing effect on the body and spirit and fulfills a need to be close to nature which is rooted deep within our human heritage.

As twentieth-century technological people we have lost our roots in nature. We have forgotten the delicate ecological web of which we and other living things are a part. Urban man has lost the sensitivity and serenity that comes with living in natural surroundings. This insensitivity to the world around us has led to widespread environmental damage. Backpacking allows one to obtain a new outlook on the world and heightens one's awareness of our dependency on nature.

Backpacking can be physically demanding; carrying a forty-pound pack along mountain trails is no easy task. However, each backpacker determines the rigors of his/her journey, as a short stroll through a forested hollow is less strenuous than a twenty-mile trek across rocky terrain. Safe backpacking necessitates an understanding of one's limitations and can be enjoyed by everyone, including children.

Living simply in nature and carrying everything on one's back requires care in selecting the proper equipment. For a detailed analysis of equipment and how to use it, consult *Backpackers Magazine,* or read Colin Fletcher's *Complete Walker*. The following suggestions may be helpful:

BACKPACKING BASICS

- Boots–Boots must be firm to protect the feet and ankles from the heaviest load one expects to carry. Popular soles today are heavily lugged, rubber-synthetic-compound types such as Vibram. Boots should be light (3 to 5 pounds) without compromising ruggedness. GI tropical boots with heavy-duty soles and light canvas tops are excellent for three-season hiking and are inexpensive.
- Pack–Two types of packs are commonly used: 1) The lightweight aluminum frame and pack bag, and 2) the frameless rucksack. For trips longer than two days, an aluminum frame and pack are required. These packs are angled at the shoulder and waist to fit the body's contours; nylon bands rest against the back (see diagram). Straps from the bottom part of the frame fasten below the waist, placing the weight of the pack on the hips. When the waist strap is released, the frame hugs the back, which allows more stability in difficult terrain. When buying a pack, keep in mind two factors besides price: 1) the intended use, and 2) the length of the trip. Look for quality items on sale at end-of-season closeouts, or find someone who is selling equipment.

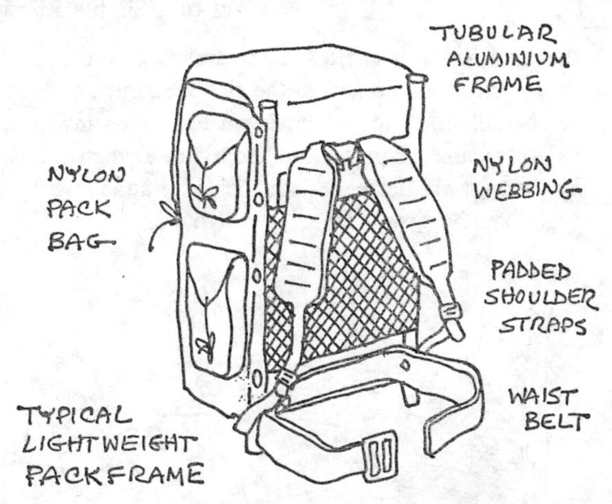

- Sleeping bag–Mummy bags are the most efficient type for conserving body heat. The type of "fill" should be chosen according to personal needs. Two high-quality fills available today are down (goose or duck) and fiberfill. Down is useful in cold climates and is the lightest material available. Fiberfill is synthetic, about one-third heavier than down, and is preferred for use in wet climates. It can be easily dried and retains its warmth when soaked; down does not. When buying a sleeping bag, assess personal needs and get the best quality within one's budget.

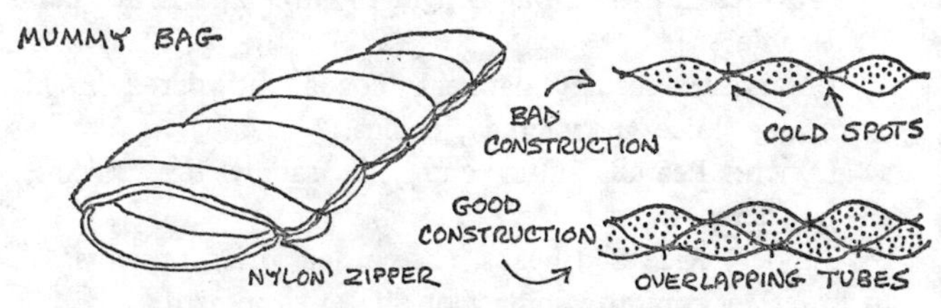

- Shelter–A bed beneath the stars has aesthetic appeal, but it is always wise to carry some type of shelter. Although the best form of shelter, tents are not always needed. A tent closes one off from the sights, sounds, and smells of the forest. Also, a tent is an expensive investment. Other means of protection from the elements include a fly sheet, plastic sheet with Visklamp attachment, tube tent, poncho, groundsheet, or bivouac.

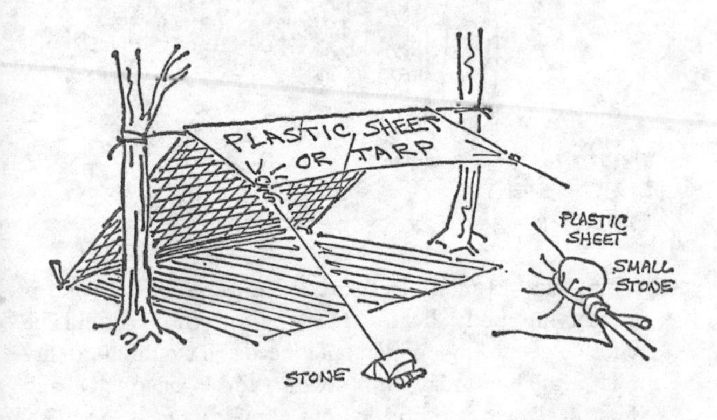

- Other Gear–Rain gear, ground pad, flashlight with extra batteries and spare bulb, map and compass, nylon rope, cooking utensils, canteen, cup, spoon, knife, match safe, and first-aid kit. Clothing and food requirements will vary depending on the season, locale, length of trip, and individual taste. Here is a simple selection of food that provides nutrition, but is light to carry and is not perishable:

  Breakfast: Instant oatmeal, cocoa, powdered milk, orange juice (powdered), raisins.

  Lunch: Peanut butter and jelly sandwiches, raisins, powdered milk.

  Supper: Rice and beans, cheese, tea, dried apricots.

In order to minimize impact on the environment:

- Carry all non-burnable trash.
- Burn paper in the campfire where permissible.
- Bury food scraps in the ground.
- Always carry a light shovel to dispose of human excre-

ment. Dig a hole 6 inches deep at least 50 feet from the nearest water. After the deposit is made, cover with loose soil and sod.

HYPOTHERMIA

A number of backpacking accidents and fatalities can be caused by hypothermia–a loss of body heat that leads to exhaustion, confusion, unconsciousness, and death. It can strike in any season and any climate. People who are new to outdoor life are especially susceptible to the effects of wind, rain, and cold. Conditions leading to hypothermia include:

- Improper clothing–Overdressing can produce sweat soaked clothing, which is as bad as being caught in a rainstorm. Not wearing enough clothing is just as dangerous. Wear several layers of thin clothing, so that they can be easily removed when exercising and put back on when chilled. Wear wool instead of cotton outer clothing; wool retains its warmth when wet, cotton does not.
- Exhaustion–Do not push too hard when hiking; allow time for rest.
- Skipping meals–Do not skimp on food because the body needs fuel for warmth; eat small meals often.
- Wind–Strong winds can quickly lower body temperature, especially when clothes are wet from rain or perspiration. Always seek shelter when a storm approaches.

Hypothermia develops in stages. As the body is chilled, it shivers as a warming response. If dizziness, confusion, and uncontrolled shivering set in, the body temperature has dropped significantly and death can quickly follow.

If one feels hypothermia's symptoms approaching, find a warm shelter and drink hot fluids. In wilderness surroundings, strip the victim and sandwich him/her between two bare bodies in a sleeping bag or blankets. This should quickly raise the body temperature.

Reference

The Complete Walker, by Colin Fletcher, Alfred A. Knopf, Inc., New York, NY, 1974.

Additional Sources

Backpacker Magazine, 28 W. 44th St., New York, NY 10036.

The Sierra Club Wilderness Handbook, by David Brower, Ballantine Books, New York, NY, 1968.

Backpacking, Tenting and Trailering, by Blackwell S. Duncan, Rand McNally, New York, NY, 1975.

Walking in the Wild, by Robert Kelsey, Funk & Wagnalls, New York, NY, 1973.

Wilderness Handbook, by Paul Petzoldt, W. W. Norton & Co., New York, NY.

Backcountry Camping, by Bill Riviere, Dolphin Books, New York, NY, 1972.

On the Loose, by Terry and Renny Russell, Ballantine Books, New York, NY, 1974.

Backpackers Digest, by C. R. Learn, Follet Publishing Co., Chicago, IL.

Backpacking for Fun, by Thomas Winnet, Wilderness Press, Berkeley, CA, 1972.

Pleasure Packing, by Robert S. Wood, Condor Books, Berkeley, CA, 1972.

The New Way of the Wilderness, by Calvin Rutstrum, Macmillan and Co., New York, NY, 1966.

Bushcraft: A Serious Guide to Survival and Camping, by Richard Graves, Schocken Books, Inc., New York, NY, 1972.

Camping and Woodcraft, by Horace Kaphart, Macmillan and Co., New York, NY, 1948.

Snow Camping and Mountaineering, by Edward Rossit, Funk & Wagnalls, distributed by Thomas Y. Crowell Co., New York, NY, 1970.

Being Your Own Doctor, by Dr. E. Russel Kodet and Bradford Angier, Stackpole Books, Harrisburg, PA, 1968.

Cooking for Camp and Trail, by Hasse Bunnelle, and Shirley Sarvis, Sierra Club, San Francisco, CA, 1972.

How to Stay Alive in the Woods, Bradford Angier, Macmillan and Co., New York, NY, 1966.

Appalachian Mountain Club (National Club)
1718 N St., NW
Washington, DC 20036

Sierra Club
324 C St., SE
Washington, DC 20003

Audubon Naturalist Society
8940 Jones Mill Rd.
Washington, DC 20015

Outward Bound, Inc.
165 West Putnam Ave.
Greenwich, CT 06830

National Outdoor Leadership School
Box AA
Lander, WY 82520

#63 ENJOY RECREATIONAL SPORTS

Two categories of recreational sports are evolving in America: 1) Active sports where the individual is physically involved and little fuel energy is required; 2) Passive sports where the individual is sedentary and considerable energy expenditures are necessary. In an energy-conscious society one should maximize the use of human resources and minimize the use of fuel energy and natural resources.

A sport like motorboating is energy intensive because it requires fossil fuel to propel the boat and to transport it to water sites. In addition, a large amount of energy goes into making the boat and engine, and into building and maintaining marinas. Compare these expenditures with swimming at the nearest lake or beach. Not only is swimming far less energy intensive, but it also provides excellent exercise and has no adverse environmental effects, as does motorboating (water, noise, air pollution, and destruction of habitat). Canoeing and sailing also demand fewer natural resources and are excellent alternatives to motorboating.

Winter sports such as skating, cross-country skiing, and snowshoeing are invigorating ways to enjoy the serenity of new-fallen snow, rather than driving a gas-guzzling snowmobile. As with motorboats, the individual participating in snowmobiling is almost completely deprived of physical activity and exercise. The snowmobile also presents problems for local inhabitants who are concerned about high noise levels. In addition, many plants and animals are damaged by snowmobile activity.

Shun energy-intensive sports that require sophisticated equipment and/or depend directly on fuel energy for enjoyment. Particularly wasteful vehicles that one could easily eliminate are golf carts and dunebuggies. Encourage sporting activities that require small energy expenditures while providing a good opportunity for exercise.

SPORTS AND ENERGY

Sport	*Equipment*	*Travel to Site*[1]	*Energy*[2]
Tennis			
(outdoors)	low	low	low
(indoors)	low	low	moderate
Swimming			
(outdoors)	nil	low-moderate	nil-low
(indoors)	nil	low-moderate	high
Football	low-moderate	low	nil
Horseback riding	moderate	moderate	moderate-high[3]
Backpacking	low-moderate	nil-moderate	nil
Basketball			
(outdoors)	low	nil-low	nil
(indoors)	low	nil-low	moderate

Sport	*Equipment*	*Travel to Site*[1]	*Energy*[2]
Soccer	low	nil-low	nil
Handball			
(outdoors)	low	nil-moderate	nil
(indoors)	low	nil-moderate	moderate
Baseball	low	nil-low	nil
Snowmobile	high	high	high
Boating			
(motor)	high	moderate	high
Fishing			
(row boat)	low	moderate	nil
(deep sea)	high	high	very high
Golf	high[4]	moderate	high[5]
Skating			
(roller)	low	low	nil
(ice, indoor)	low	low	moderate
Bowling	high	low	high
Skiing	high[6]	high	high
Hiking	nil	low	nil
Jogging	nil	nil	nil
Parachuting	high	high	very high
Judo, gymnastics	low	low	moderate
Sailing	moderate	moderate	nil
Race car driving	high	moderate	very high

Make better use of open space by making it a recreational center for young and old alike. A few simple hints are:[7]

- Toddler's play place–requires a sandbox, teeter, and a swing.
- Box hockey–make a plywood box 2′ by 6′ by 5″, with a center divider notched at the top center and on either side at the bottom. A 2″ by 1″ by ¾″ wooden puck

is knocked with a broom handle through the lower notches.

- Tether ball–one needs a space 20 feet in diameter. A pole 10′ above the ground and 2½″ in diameter has a rope 7–8′ long attached to the top. The other end is tied to a bag containing a tennis ball. The aim is to hit the bag with a wooden paddle and wind it around the pole; the opponent hits to prevent winding.
- Deck tennis–one needs an area 24 by 50 feet, a net and rubber rings.
- Paddle tennis–play like tennis with wooden paddles and a rubber ball.
- Clock golf–sink tin cans without ends into the lawn; use a putter to knock golf balls into the cups.
- Badminton–requires an area 24 by 54 feet.
- Shuffleboard–one needs a concrete slab 6 by 45 feet.
- Horseshoe pitching–requires a narrow area 30 to 40 feet long.
- Bowling on the green–one must have a smooth level lawn at least 10 by 50 feet and croquet balls.
- Croquet–requires an area 30 by 60 feet, balls, mallets, and hoops.
- Volleyball–one needs an area 40 by 70 feet.

Notes

1. Average travel by urban Americans.
2. Fossil fuel expenditure, not human energy. It includes fuel for heating indoor facilities and apportioned according to number of users.
3. Production of grain and hay for horse included.
4. Includes golf cart.
5. Large expenditure of fertilizer for grounds.
6. Includes ski lift.
7. *A Guide to Home Landscaping,* Donald J. Bushey, McGraw-Hill, Inc., New York, NY, 1956. (Reprinted with permission of McGraw-Hill, Inc.)

Additional Sources

New York Times Book of Home Landscaping, edited by Joan L. Faust, Knopf Publishers, New York, NY, 1964.

Wilderness Canoeing, by John Malo, Macmillan and Co., New York, NY, 1972.

Swimming Techniques in Pictures, by Bob Horn, Grosset & Dunlap, New York, NY, 1974.
Athletic Fitness, by Dewey Schurman, Atheneum Press, New York, NY, 1975.
Soccer Illustrated: For Coach and Player, by Frank F. DiClemente, Ronald Press Co., New York, NY, 1968.

#64 EXERCISE WITHOUT EQUIPMENT

The best type of exercise, requiring no fuel energy, capital, maintenance, expense, or rigorous training, is walking. Get physically fit, lose weight, and save gasoline by leaving the car behind and going on foot. Swing the arms, look at the surroundings, and greet the neighbors. Be sure to wear comfortable and sturdy shoes.

Run errands on foot instead of using the car. Walk to the grocery store and pull the purchases home in a cart. Many residential areas have a library and post office within walking distance. Wake up earlier and go on foot to work, or at least part way, and catch a bus the rest of the way. Most people find they can go longer distances on foot than they thought and will gladly leave those car keys at home. During the holiday season, ring in the spirit by caroling and delivering presents on foot.

Group-walking activities combine exercise and companionship. Hiking clubs and bird watchers enjoy the outdoors as they exercise. Mountain climbing requires equipment and courage, but it builds strong legs and good times.

Jogging has special advantages for adults as it helps develop the cardiovascular and respiratory systems. A simple exercise, it requires no skills, little time, and can be done anywhere outdoors. The degree of exertion is controlled by the individual, who alternately walks and runs. The run is steady and easy paced, and can alternate with a breath-catching period of walking. For the beginner, the run is a slow regular trot, one pace up from walking.

Jogging's benefits are multiple: it slims the body; sagging muscles are firmed; aches and pains are reduced; heart and lung efficiency is improved; legs are strengthened.

For best results when starting to jog regularly, heed the advice of experts:

- Those over thirty should first have a medical checkup.
- Stand up straight with the head up.
- Hold elbows slightly bent and away from the body.
- Move legs freely from the hips with an easy motion.
- Relax the ankles; land first on the heel, then rock forward to the toes.
- Inhale through the nose and exhale through the mouth.
- Dress for the season. Wear proper shoes with a thick rubber, crepe, ripple, or neolite sole.
- Keep a record of progress.

In the past, vigorous activity was a part of daily life. Today, mechanical devices make living so comfortable that most of us do not get enough exercise to stay physically fit.

Fitness is more than simply building strength to perform daily tasks. It also provides:

- Endurance by reducing fatigue, especially from mental labor.
- The ability to absorb stress.

- Greater resistance to illness.
- Improved recuperative power.
- Optimum on-the-job efficiency.
- A more enjoyable life.

Additional Sources

An Introduction to Physical Fitness, a booklet by The President's Council on Physical Fitness and Sports, Superintendent of Documents, U.S.G.P.O., Washington, DC 20402.

The New Aerobics, by Kenneth H. Cooper, Bantam Books, Inc., New York, NY, 1970.

Creative Walking for Physical Fitness, by Harry J. Johnson, M.D., Grosset & Dunlap, New York, NY, 1970.

Frisbee, by Stancil E. Johnson, Workman Publishers, New York, NY, 1975.

VIII HEALTH

INTRODUCTION

Although the United States is one of the wealthiest and most powerful nations on earth, our infant mortality rate ranks fifteenth in the world, male life expectancy nineteenth, and female life expectancy seventh.[1]

The tensions of American life cause mental and nervous disorders and have created a society that depends on over-the-counter drugs to pep up and calm down. The competitive pressures to succeed and to line our pockets with gold has dulled our sensitivities to life's natural beauties and simplicities.

Tips on maintaining proper health should not be regarded as strict do's and don'ts, but rather as suggestions to simplify life and to gain more fulfillment. Everyone should analyze the use of personal products and decide if these items are really essential for health and hygiene.

We know the dangers of abusing tobacco, alcohol, and the unhealthy aspects of overeating and failing to exercise, and we know that built-up pressures must be released. But it is the individual's responsibility to use common sense in these matters. A healthy body is a precious gift and should not be neglected.

Note

1. *1973 Demographic Yearbook of the United Nations,* United Nations Statistical Office, International Publications Service, New York, NY, 1973.

#65 USE SIMPLE PERSONAL PRODUCTS

With slick advertising and psychological gambits, the cosmetics industry has convinced most Americans that they must attain the ideal body, breath, fragrance, and personality. Cosmetics are even required to attain a natural look.

Over eight hundred manufacturers with gross sales of more than $8 billion per year flood the market with products needed for the ideal appearance. However, it is possible to be attractive without the expense of using a variety of products. Many products are not really necessary, while others might be needed only on special occasions.

Deodorants and antiperspirants, among the highest sales items, might be needed in hot weather, for special events, or with certain clothing, but for most people a daily shower and fresh clothing is enough to prevent body odor. If one uses a deodorant, use a stick, a roll-on, or baking soda, rather than an aerosol spray.

Women's fashion trends have dictated that underarms and legs must be free of hair. Women should exercise personal preference and let it grow if they so desire. Men who grow beards gain additional winter warmth and save on the cost of equipment, time, and energy used by electric shavers and shaving blades.

TO SUBSTITUTE FOR:	WHICH COSTS	MIX AND USE	WHICH COSTS	AND SAVE
Ban Basic Deodorant	$1.97/3 oz.	1 tablespoon alum 1 cup water	$.03/oz	$1.94
Arm & Hammer Baking Soda Deodorant	$1.74/7 oz	3/4 cup cornstarch 1/4 cup baking soda	$.19/7 oz	$1.55
Listerine Mouthwash	$1.09/20 oz	1/2 teaspoon salt 1 cup water (several drops of peppermint oil extract for flavor	$.02/20 oz	$1.05
Mennen Lime Electric Preshave	$1.51/6 oz	Talcum powder or Cornstarch	$.50/6 oz. $.16/6 oz	$1.01 $1.35
Rapid Shave Aerosol	$1.19/11 oz	Brush & mug and Soap	$.66/4 oz	$.60
Liquid Styptic Pencil		Use the alum solution above		
Crest Tooth Paste	$1.03/7 oz	Baking soda	$.24/7 oz	$.79
Cutex Herbal Fingernail Polish Remover	$.59/4 oz	8 oz acetone 2 oz water 1 teaspoon olive oil	$.30/3 oz	$.20

Bad breath can be caused by highly seasoned food, plaque on the teeth, decayed teeth, gum diseases, and infections in the nasal passages and throat. Scented mouthwashes and lozenges only mask the odor. The proper treatment is to remove the cause. Salt or baking soda mixed in water is an adequate mouthwash.

Everyone should keep his/her natural hair color. Have a hair style that does not require a lacquer coating to keep it in place.

The homemade products on page 247 work as well or better than fancy, expensive commercial products. The basic chemicals used are available in local supermarkets and drugstores.

Reference

"Cosmetics From the Kitchen," by Marcia Donnan in *Consumer's Survival Kit,* Praeger Publishers, New York, NY, 1965.

Additional Sources

Making Soap for Gifts and Personal Use, by Ann S. Bramson, Workman Publishing Co., New York, NY, 1975.

Clinical Toxicology of Chemical Products, by Mary Ann Gleason, et al., Williams and Wilkins, Baltimore, MD, 1969.

Chemical Formulary, by Harriet Bennett, Chemical Publishing Co., New York, NY, 1975.

#66 SELECT SAFE COSMETICS

Consumers who suffer adverse reactions from cosmetic products do not always voice their complaints. Some merely suffer in silence and perhaps stop using the product. Others return the product to the seller for a refund.

If serious side effects result from using a certain cosmetic product, notify the Consumer Product Safety Commission, the Food and Drug Administration (FDA), and the product manufacturer. If the injury is serious, see a doctor.

A 1974 study of cosmetics injuries, conducted by the American Academy of Dermatology and the Food and Drug Administration, revealed that most complaints are from people with minor skin irritations caused by deodorants, antiperspirants, and hair removers. Skin moisturizers and hair sprays also caused a significant number of injuries.

When using cosmetics, follow the directions and warnings for the intended use. Patch test new products on a small area of skin. Basic cleanliness rules should be followed—wash hands before applying; keep containers closed when not in use; never use another person's cosmetics; and never substitute saliva for water (especially for eye cosmetics).

All personal products packaged as aerosol sprays should be avoided. The propellant can cause skin irritation and the ingredients can be easily breathed and absorbed into the blood stream (see entry #17).

The feminine deodorant spray is one non-essential product with a high injury rate. For cleanliness, soap and water are more effective than perfumed sprays. The word "hygiene" has disappeared from the product name since these sprays have no medical usefulness. Reports of infection, itching, open sores, urinary urgency, and rashes are quite common. Since shyness prevents many women from consulting a physician about genital problems, the dangers of these products should be brought to the public's attention.

As of early 1976, federal regulations require that cosmetic

labels list ingredients in order of predominance, which should increase public awareness and product safety. Consumers will be more able to pinpoint ingredients that cause adverse reactions, and thereby avoid those products. One may be surprised to learn that many expensive cosmetics contain wheat flour, cocoa butter, egg, vegetable oil, and other simple ingredients —which could provide incentive to produce homemade beauty aids.

References

An Overview of Cosmetic Regulation, by Robert Schaffner, F.D.A. Report, Superintendent of Documents, U.S.G.P.O., Washington, DC 20402, April 1972.

"Aerosol Sprays," by Barbara Hogan and Dennis Darcey, CSPI Publications, 1757 S St., Washington, DC 20009, 1976.

"A Revolution in Cosmetic Regulation," by Jane Heenan, FDA Consumer Bulletin, Superintendent of Documents, U.S.G.P.O., Washington, DC 20402, April 1974.

Additional Sources

The Complete Book of Natural Cosmetics, by Beatrice Traven, Simon and Schuster, New York, NY, 1974.

Food and Drug Administration
5600 Fishers Lane
Rockville, MD 20852

Consumer Product Safety Commission
1750 K St., NW
Washington, DC 20207

See "Additional Sources" in #93 "Consumers Beware."

#67 BEWARE OF HAIR DYES

Personal products containing chemicals can bring serious side effects; some are immediately experienced while others remain latent for years. Cancer is one example of a disease that results from an accumulation of harmful substances over a long period of time.

Dyes contained in hair-coloring compounds can accumulate in the body to cause harm. Over 20 million American men and women alter their natural hair color. The amount of self-inflicted damage depends on the dyeing method chosen.

If using hair dyes, heed the warning that certain ingredients may cause skin irritations and test the product before using it. Hair dyes should never be used to color eyelashes or eyebrows. If they accidentally enter the eye, blindness may result.

Although the ingredients can be absorbed by the scalp, there is no law requiring that hair-dye chemicals be tested for carcinogens or mutagens. The existing evidence suggests that some hair dyes may cause cancer. If using hair dyes, beware—they may be hazardous to health. Remember, a simple life means keeping things natural, including one's hair.

HAIR-DYEING METHODS[1]

- Rinses—The dye does not penetrate the hair shaft; they are water-resistant but are removed in shampooing.
- Semipermanent treatments—Dyes are used to blend streaked hair, color white hair, or add highlights. The colorings are mild. The dye penetrates the hair shaft but is washed out in about four shampoos.

"Permanent" Coloring:

- Vegetable dyes—Henna is one that has been used for centuries but is not currently popular; it gives an unnatural orange-red color and makes the hair stiff and brittle.
- Metallic dyes—Lead acetate and, occasionally, silver and bismuth compounds react chemically on the hair for

a coating of metal sulphide pigments. If the scalp is cut, lead can be absorbed into the body.

- Oxidation dyes–These are the most commonly used. Ammonia-soap in the formula swells the shaft; detergents help the peroxide bleach and the dye to penetrate the shaft.

Note

1. "If You're Coloring Your Hair," by Jane Heenan, reprint from *F.D.A. Consumer*, November 1974.

Additional Sources

The Toxic Substances List-1975, edited by Herbert Cristensen, H.E.W. Publication #1733-00096, Superintendent of Documents, U.S.G.P.O., Washington, DC 20402.

#68 DO NOT ABUSE DRUGS

An interesting twist to the controversial "drug problem" is society's overreliance on legal drugs and over-the-counter (OTC) medications. Americans use and misuse drugs to calm, stimulate, sleep, lose weight, and escape reality. Our medical system leans heavily on drugs as temporary remedies for illnesses.

Consumers spend over $8 billion for these products,[1] some of which are not effective. Part of the drug issue can be traced to an industry that promotes pills as cures for medical and emotional afflictions. For too many people, aspirin, laxatives, antihistamines, and caffeine tablets are required to make it through the day. The national cure for tension, nervousness, and depression is a pill—instead of a change in lifestyle.

To illustrate America's psychological dependence on "medications," watch the evening news on any national network and take note of the products advertised: remedies for headache, upset stomach, constipation, insomnia, and tension. Advertisements cause one to wonder if pharmaceutical firms are trying to earn a dollar by playing on the health insecurities of millions of Americans.

Most of the newly marketed drugs are merely modifications of old products. "Fast acting" usually means that caffeine has been added. A congressional subcommittee investigating drugs found that promotion costs range up to three times that spent on research; it was estimated that $.25 of every product dollar went toward advertising costs. It was also discovered that 70 per cent of advertised drugs were not effective for the ailments they supposedly cured. Also, 65 per cent of the physicians interviewed admitted that drug industry sales personnel—not medical researchers—are the main contacts for finding new products.[2]

Aspirin is the most popular self-medication for minor pains and fever. It is the medicine most frequently involved in accidental childhood poisonings. As with every drug, it can have serious side effects and should not be given indiscriminately to everyone. People with stomach ulcers especially should avoid aspirin. In many cases, when aspirin is sought as a "cure" a few minutes of quiet relaxation would provide a remedy.

Sleeping pills are popular products with a questionable use and effectiveness. A person who has trouble sleeping should consult a physician and not rely on self-medication. Restlessness may result from drinking caffeinated coffee or tea in the evening or possibly from not being physically tired at bedtime.

Fostered by laxative advertisements, a number of misconceptions prevail concerning bowel movements. Regularity is thought to be a daily movement; however, individuals vary and a person should not expect clockwork regularity. Instead of relying on pills, eat a diet rich in fiber: fruits, vegetables, bran, whole grains. Drinking plenty of liquids and proper physical exercise should also help eliminate bowel concerns.

Cold remedies are another large seller that could be used less frequently by following common-sense rules such as staying out of crowds, getting plenty of sleep, and having a nutritious diet.

Proper nutrition and exercise could significantly reduce our national health bill and result in a more cheerful population that is less dependent on drugs.

Drug Safety Tips[3]

- Avoid a dependency on any remedy that "cures" real or imagined ills, since all drugs affect the body's chemistry.
- Use OTC's only when absolutely necessary and for short-term ailments.
- Use prescriptions only on a doctor's orders. Do not take a prescription written for someone else.
- Date all medications when purchased.
- Buy only in small quantities.
- Store out of the reach of children.
- Follow directions carefully.
- Do not take medication from an unlabeled bottle.
- Do not combine medications.
- Weed out leftovers and old medicines regularly.
- Discard drugs carefully, and make sure that children cannot salvage them.

Notes

1. "National Health Expenditures by Object: 1950–1972," *The US Factbook–1975;* 95th edition, Grosset & Dunlap, New York, NY, October 1974.

2. "Kennedy Subcommittee Hearings on the Pharmaceutical Industry," from *Organizing for Health Care, Source Catalog 3,* Beacon Press, Boston, MA, 1974.
3. "First Facts About Drugs," a pamphlet by the Public Health Service and the Food and Drug Administration, Superintendent of Documents, U.S.G.P.O., Washington, DC 20402, November 1971.

Additional Sources

Drugs: Administering Catastrophe, by Graham S. Finney, Drug Abuse Council, 1828 L St., NW, Washington, DC 20036, 1975.

"Education and Drugs," from *Edcentric,* P.O. 10085, Eugene, OR 97401, March 1972.

The American Connection: Politicking and Profiteering in the Ethical Drug Industry, by John Pekkanen, Follet Publishing Co., Chicago, IL, 1973.

The Great Drug Deception, by Rolf Adam Fine, Stein and Day, New York, NY, 1970.

The New Handbook of Prescription Drugs, by Richard Barack, M.D., Ballantine Books, New York, NY, 1970.

Mystification and Drug Misuse, by Henry L. Lennard Associates, Jossey-Bass Inc., San Francisco, CA, 1971.

National Coordinating Council on Drug Education: Suite 212, 1211 Connecticut Ave., NW, Washington, DC 20036.

Drug Advertising Project: National Council of Churches, 100 Maryland Ave., NE, Washington, DC 20002. (Investigate false advertising claims and promotion of legal drugs.)

#69 STOP SMOKING CIGARETTES

"Warning: The Surgeon General Has Determined That Cigarette Smoking May Be Dangerous to Your Health." In 1965 this caution was placed on all cigarette packages and in advertisements.[1] Twenty-nine million ex-smokers attest to the warning's effectiveness and the increasing knowledge of smoking's hazards, and the 43 per cent of adults who smoked in 1965 dropped to 36 per cent by 1971. Unfortunately, the numbers are increasing again, as more women and young people are smoking. The health hazards are numerous, well documented, and should be sufficient to make people kick the habit for life.

On the average, smokers die at a younger age than non-smokers. A person who smokes one-half pack a day surrenders five and one-half years of life, while a pack-a-day person pours a cup of tar into his/her lungs every year. Physiologically, smoke paralyzes respiratory cleaning mechanisms and subjects the lungs to invasion by bacteria and carcinogens.

Lung cancer is the second most frequent cause of death in smokers, and 90 per cent of lung cancer patients are smokers. An early diagnosis is difficult to make; therefore, surgery is rarely performed while the cancer is small. Only 10 per cent of lung cancer patients are saved.

Smokers are more prone than non-smokers to develop cancer in other body areas, such as the mouth, larynx, esophagus, and urinary tract. Smoking is one of the four major risk factors of heart attacks (along with high blood pressure, obesity, and high blood cholesterol). Smokers have 70 per cent more heart attacks than non-smokers and also suffer a high number of strokes. Women who smoke during pregnancy have more stillbirths, first-month infant deaths, and are more apt to have underweight babies at birth.

Deaths from emphysema and chronic bronchitis increased by 1,000 per cent between 1945 and 1968, and most of the dead were smokers. Smokers are absent from work more days of the year than non-smokers. Smoking is a serious indoor air pollutant and can irritate others in the same room.

Cigarette smoking is also a fire hazard responsible for thousands of deaths and millions of dollars worth of property damage each year. Over 20 per cent of all fires in the United States are caused by smoldering cigarettes or by the matches used to light up.

To grow, cure, process, package, and transport tobacco products such as cigarettes requires a large expenditure of fertilizer and fossil fuel energy. The sad irony is that these resources could be utilized in agricultural production to feed the starving people of the world rather than to produce a product that is harmful to human health.

TIPS TO STOP SMOKING[2]

- Smoke one less cigarette each day.
- Make each cigarette a special decision–and keep putting it off.
- Don't give up cigarettes *completely*. Carry *one* in case of need and save it–permanently.
- Don't quit "forever," just stop for one day. Tomorrow, try quitting a second day, tomorrow, a third . . .
- Tell friends and family of intentions to quit. A public commitment will boost willpower.
- Hide evidence–cigarettes, ashtrays, and matches.
- Have a supply of chewing gum, carrot sticks, celery, nuts, etc.
- Nervousness and hunger indicate the body's readjustment; if too upsetting, ask a doctor for help.

If one refuses to stop, reduce the danger:[3]

- Choose a brand with low tar and nicotine. Filtering and reducing the amount of tobacco lowers tar and nicotine.
- Don't smoke an entire cigarette.
- Take fewer draws on each cigarette.
- Take short shallow puffs instead of inhaling.
- Smoke fewer cigarettes each day.

Notes

1. On April 1, 1970, this warning was changed to read, "Warning: The Surgeon General Has Determined that Cigarette Smoking Is Dangerous to Your Health."
2. "Danger–Cigarettes," American Cancer Society, 219 East 42nd St., New York, NY 10017.
3. "If You Must Smoke," a U.S. Department of Health, Education and Welfare leaflet, Superintendent of Documents, U.S.G.P.O., Washington, DC 20402.

Additional Sources

"75 Cancer Facts & Figures," American Cancer Society, 219 East 42nd St., New York, NY 10017.

"Tar and Nicotine Content of Cigarettes," a Federal Trade Commission leaflet, Superintendent of Documents, U.S.G.P.O., Washington, DC 20402, March 1974.

"Women and Smoking," by Jane E. Brody and Richard Enquist, Public Affairs Pamphlet #475, 381 Park Ave. South, New York, NY 10016, 1972.

Action on Smoking and Health
2000 H St., NW
Washington, DC 20006

#70 CURB ALCOHOL ABUSE

Alcohol can be classified as a food or a drug, depending on how one uses it. Almost 70 per cent of America's adults drink occasionally. Throughout history, alcohol has been used by people of many cultures for various social and religious reasons, at celebrations, to relax, or to complement dinner. Most people drink in conjunction with other activities, and when it is part of a social or religious custom, there is little problem drinking.

On the other hand, it is estimated that in America, one drinker out of ten abuses alcohol. People who have difficulty facing daily activities drink to escape, forget worries, gain courage, or calm down.

Ethyl (drinking) alcohol is formed by the fermentation of sugar with yeast. Different beverages are made by using different sugar sources in fermentation—beer uses germinated barley; wine uses grapes or other fruits, whiskey uses malted grains; rum uses molasses. The food resources and energy consumed in the production, processing and packaging of alcoholic beverages are quite substantial. Beer, for example, requires about 35,000 Btu/gallon to produce, wine 28,000 Btu/gallon and hard liquor about 85,000 Btu/gallon.[1]

When drinking, about 20 per cent of the alcohol is absorbed directly into the blood through the stomach. The blood carries the alcohol to the brain where it slows mental activity and depresses the central nervous system.

Initially, alcohol stimulates the drinker as it affects the individual's self-control. Inhibitions may temporarily disappear and the person may feel aggressive, depressed, or energetic. Increased intake may affect brain activity so that memory, muscular co-ordination, and judgment may be adversely affected.

The way that alcohol affects a person depends on a number of factors. If one drinks slowly, not more than one drink per hour, the alcohol will not jolt the brain or build up in the bloodstream. However, gulping will usually bring rapid intoxication.

The amount of food in one's system is another consideration. Eating while drinking will slow the rate of absorption. The kind of drink is another factor, with wine and beer being less potent. Diluting drinks with water will slow absorption but mixing with carbonated beverages will speed the effects. A drinker's weight plays a role in that a 120-pound person will have less tolerance than a 180-pound person.

Because alcohol affects the brain, it does have a potential for danger. Limited intake may not damage the body or brain, but overconsumption will cause serious health problems. Drinking abuse has been associated with ulcers, heart disease, cirrhosis of the liver, and severe mental and nervous disorders. Drunkenness is responsible for illness, crime, and accidents. About 100,000 deaths a year in the United States are related to alcohol. Fifty per cent of homicides and felonies can be linked with drinking.[2] One third of all suicides show signs of alcoholism. The disease is involved in a majority of accidental falls, drownings, and fire-related deaths. Roughly 131,000 veterans discharged from veteran's hospitals and 25–50 per cent of the total hospital population have drinking problems. The National Highway Traffic Safety Administration associates 40 per cent of all highway fatalities with alcohol.[3]

Much too frequently, drunkenness is seen as a humorous way of acting, which may play a part in America's drinking problems.

Drinkers should remember that one cannot sober up with black coffee, a cold shower, or fresh air. Alcohol takes time to burn off; generally, one should allow one hour of sobering time for each drink consumed.

Drink responsibly to minimize problems–sip slowly, eat while drinking, and drink under relaxed social conditions. Everyone should know the signs of alcohol abuse:

- The need to drink before social situations.
- Frequent intoxication.
- An increase in the amount consumed.
- Drinking alone.
- Drinking in the early morning.
- Drinking to avoid problems.

Everyone interested in human kindness and quality lifestyle should show concern for those who abuse alcohol. There are an estimated 10 million alcoholics in America–and very few of these are the skid-row bums that people sometimes associate with problem drinking. If a family member or friend has a drinking problem, help him/her by contacting local chapters of Alcoholics Anonymous for assistance.

Finally, remember that drinking is expensive. Americans spent $21.5 billion on alcoholic beverages in 1974, or $100 for every man, woman, and child.

Notes

1. *Energy and Food,* by Albert J. Fritsch, Linda Dujack, and Douglas Jimerson, CSPI Energy Series VI, CSPI Publications, 1757 S St., NW, Washington, DC 20009, p. 59.
2. "Medical Complications of Alcohol Abuse," a pamphlet by the Committee on Alcohol and Drug Dependence, AMA, 535 N. Dearborn St., Chicago, IL 60611.
3. "Social Cost of Drug Abuse," by Special Action Office for Drug Abuse Prevention, Washington, DC, December 1974.

Additional Sources

Organizing for Health Care, Source Catalog 3, Beacon Press, Boston, MA, 1974.

Alcoholism, a series of pamphlets prepared by the National Institute of Mental Health, Superintendent of Documents, U.S.G.P.O., Washington, DC 20402.

National Clearinghouse on Alcoholic Abuse
Rockville, MD 20852

National Council on Alcoholism
2 Park Ave.
New York, NY 10016

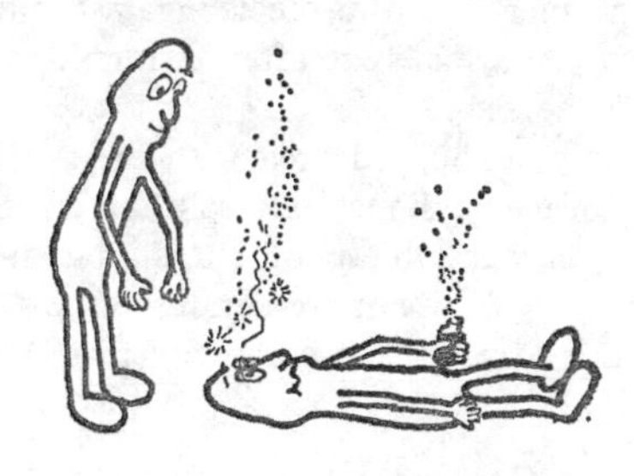

#71 PREPARE FOR ENVIRONMENTAL EXTREMES

Certain precautions should be followed in times of severe weather conditions such as extreme heat, cold, or pollution.

Avoid Sunburn

The concept of a tanned body as an ideal physical condition is somewhat misleading, since the body's tan is a protective reaction to the sun's rays. Sunshine can produce vitamin D in the skin, but one doesn't have to roast to get it, and, besides, this vitamin is available in fortified milk and other foods. Light-skinned people are especially prone to skin cancer from ultra-violet rays.

Lengthy exposure can cause bad burns with concurrent

pain, blisters, and sometimes chills and fever. Too much tanning speeds the aging process as the dried skin becomes leathery and wrinkled. Tan gradually to allow the skin's protective ingredient, melanin, to be utilized more effectively.

General Tips for Tanners[1]

- People with average sensitive skin can stay in the sun about twenty minutes the first day and increase the time by five minutes each following day. When a tan is developed, one can stay in the sun most of the day without burning.
- The elderly, small children, and people with bald spots should be especially careful about sunburn.
- A sun-screening lotion protects against burning, but it won't stay on the skin after swimming, toweling, or lying on the sand.
- Dust a mildly sunburned area with cornstarch.
- A dressing of gauze dipped in a solution of one tablespoon baking soda, one tablespoon cornstarch, and two quarts of water will ease a more serious burn.
- If the burn is severe, with chills, fever, and blisters, see a doctor.

Heat-wave Safety Rules[2]

- Slow down; the body does not function best in high temperatures and humidity. If affected by heat, go to a cooler area and reduce activity.
- Dress for the summer with lightweight and light-colored clothing, which reflects heat and sunshine.
- Eat less food. Foods that increase the metabolic heat process also increase water loss.
- Do not become dehydrated. Drink plenty of water and add an occasional salt tablet to the water if one is active during a heat wave.
- Avoid thermal shock by staying out of the heat for a few hours each day.
- Don't sunburn—this makes the job of heat dissipation more difficult.

Protection from Air Pollution

- Keep the house as dust free as possible.

- Use household air filters and change them often.
- Avoid machinery that emits fumes.
- If allergic, stay away from feather-filled bedding, lint-producing articles, or any object or animal to which one might be sensitive.
- Prohibit tobacco smoking.
- If one has lung problems and lives in an area with high-pollution levels, consider moving.
- When the air quality is at a hazardous level, stay indoors, especially if one has lung or heart difficulties.
- Do not use aerosol sprays in the home.

Cold-weather Safety Rules

- Stay away from cold drafts.
- Keep warm and dry—wear proper clothing.
- Drink hot fluids and eat hot meals.
- Take care of chapped skin.
- Remove heavy outer clothing when indoors, on the bus, or in a warm place.
- Cover the face in extremely cold weather.

Hints for Avoiding Frostbite

- Cover the nose, cheeks, ears, toes, and fingers with warm clothing.
- If frostbite does occur, do not rub the area (usually dead white in color), and do not expose it to high temperatures immediately. Bring the victim into a warm room and give him/her a warm drink.
- The frozen part should be handled carefully, immersed in lukewarm water, or gently wrapped in warm blankets.
- When the frost-bitten part is rewarmed, the patient should be encouraged to exercise the affected area. If the condition is severe, take the person to a medical facility as quickly as possible.

Notes

1. "In Pursuit of a Summer Tan," by Jane Heenan, *FDA Consumer Newsletter,* May 1974.
2. "Heat Wave Safety Rules," from Public Works, edited by Walter Szykitka, Links Books, New York, NY, p. 108.

Additional Sources

"Winter Storms," from the National Weather Service, Superintendent of Documents, U.S.G.P.O., Washington, DC 20402.

"Heat Wave," from the National Weather Service, Superintendent of Documents, U.S.G.P.O., Washington, DC 20402.

Accidental Death and Disability: The Neglected Disease of Modern Society, Division of Medical Sciences, National Research Council, 2106 Constitution Ave., NW, Washington, DC 20418, 1970 (free).

#72 GUARD AGAINST HOME HAZARDS

The home is a place where frequent accidents can occur. Preventing unnecessary injury to family members should take high priority.

Home Fire Escape Plans:[1]

- Have at least two ways out of each room.
- Close bedroom and hall doors at night; fire can sweep up stairs and through halls with alarming speed.
- Use a window as a possible escape route; be sure that children know how to open it.
- If trapped in a bedroom, keep the door closed; open a window at the top and bottom for fresh air and hang a sheet out to signal rescuers.

- Plan a simple family alarm signal–a whistle, pounding on walls, or simply shouting.
- Get out fast; don't dress, gather valuables, or fight the fire.
- Don't open a door if the panels are hot or if smoke is leaking around the edges. Be ready to slam the door if heat and smoke rush in when it is opened.
- Notify the fire department immediately.
- Speak slowly and distinctively when talking with the fire department; give the name, address, and answer questions. When using a street alarm box, stay at the box to direct the firemen to the house.
- Have a family fire drill at least once a year.

Ways to Prevent Accidental Poisoning:[2]

- Lock cabinets containing medicines. When administering medications to children never refer to them as "candy."
- Replace torn or lost labels of medicine bottles and cover with transparent tape. If a medicine is unidentified, throw it away.
- Never leave prescription medicines around the house on tables, dressers, or in pocketbooks.
- Keep dangerous substances such as cleaning fluids, paint thinners, and poisons in their original containers with proper labels and safety instructions.
- Never store household cleaners in low cabinets or on shelves accessible to children.
- Never leave the room while using a poisonous household product. Always take it along when answering the phone or door.
- Never leave a pressurized spray container within a child's reach; never dispose of aerosols in a furnace or incinerator.
- Keep a "Poison and Overdose First Aid Chart" on the medicine cabinet door.

Ways to Protect Children[3]

- Children learn from watching. Teach them to do things safely.
- Don't leave dangerous things around for curious hands.
- Never leave a small child alone in the house.
- Know where the child is before driving the car.

- If the child is three years old or less, watch him/her every minute or set up a safe environment for play.
- Set a good example: obey all traffic regulations and use seat belts.
- Demonstrate the safe way to cross streets.
- Never drive over a box or pile of leaves—a child might be playing it in.
- Insist on good behavior in the car; no standing up. Lock the car doors and strap children in safety car seats.
- Keep a fire extinguisher in the house; be careful with cigarettes and matches.
- Screen the fireplace; never leave a toddler alone with a heater or fireplace.
- Replace electric cords and equipment when worn.
- Place the steam kettle and handles of cooking pans out of the child's reach.
- Measure medicine doses carefully.
- Discard old medicines properly to prevent children from taking them.
- Keep medicine and chemicals on a high shelf or locked up; put them away immediately after use; do not keep food near medicines.
- Teach children to never taste unidentified things they find: berries, roots, mushrooms, pills, and liquids.
- Never store chemical solutions in familiar food or beverage containers.
- Read directions on drug and chemical containers every time they are used.
- Do not use lead-based paints on children's toys.
- Keep plastic bags away from children.
- Be wary of vaporizers; a pan of water has the same effect.
- Remove door catches from old iceboxes and chests.
- Remove small bones from chicken and fish for small children.
- Blow up balloons for a young child; it might be sucked into the throat.
- Keep an eye on the child every minute when near water.
- Make sure every small pond in the area is fenced off.
- Never leave a child under two alone in the tub. Empty

wading pools after a day's use; even two or three inches of water is dangerous.
- Don't use metal utensils to get toast out of an operating toaster.
- Don't allow children in the back of a television set.

First Aid[4]

- Be calm, take command, and keep onlookers at a distance.
- Locate injuries, look for bleeding to control.
- If artificial respiration is needed, start immediately.
- Keep the patient lying down.
- Cover wounds with sterile or clean dressings.
- Look for fractures; if possible, never move a patient until splints have been applied.
- Use a proper dressing on burns.
- Place the patient carefully on a stretcher; avoid jerky movements.
- When delivering the patient to an ambulance or hospital, tell medical personnel if a tourniquet is in place and for how long.

Notes

1. "Home Fire Escape Plans and Drills," National Fire Protection Association: 470 Atlantic Ave., Boston, MA 02210.
2. *A Parent's Guide to Child Safety,* by Vincent J. Fontana, T. Y. Crowell Co., New York, NY, 1973.
3. "Child Care Preschool Years," from the Consumer Product Safety Commission, Superintendent of Documents, U.S.G.P.O., Washington, DC 20402.
4. *First Aid,* a Bureau of Mines Instruction Manual, Superintendent of Documents, U.S.G.P.O., Washington, DC 20402.

Additional Sources

Emergency Medical Guide, John Henderson, McGraw-Hill Book Co., New York, NY, 1963.

Home Medical Handbook, Russel Kodet and Bradford Angier, Association Press, New York, NY 10007.

The Well Body Book, by Mike Samuels, M.D. and Hal Benett, Random House/Bookworks, Berkeley, CA, 1973.

"A Study of Indoor Air Quality," by W. A. Cote, W. A. Wade, and J. E. Yocom, Research Corporation of New England, prepared for the U.S. Environmental Protection Agency, Superintendent of Documents, U.S.G.P.O., Washington, DC 20402.

National Fire Protection Association
470 Atlantic Ave.
Boston, MA 02210

#73 RECOGNIZE SIGNS OF CHRONIC DISEASE

Preserving human health is a very personal means of simple living. Detecting the signs of chronic disease will conserve that health.

Heart Disease[1]

Every year almost 700,000 people die and 300,000 are disabled from coronary heart disease. An increasing number of these are middle-aged. Certain physical conditions and living habits increase a person's chances of heart attack:

- High blood pressure–control it through regular checkups.
- Overweight–count calories and avoid excess weight; teach children to eat sensibly; overweight children tend to become overweight adults.
- Cholesterol and saturated fat–serve fish, poultry, and vegetable protein more often than pork, beef, or luncheon meats; when using meats, trim excess fat; use liquid vegetable oils and polyunsaturated shortenings instead of butter and cream; use skim milk and skim-milk products; eat fewer eggs; high-fat foods contribute to arterial blockage which reduces blood flow and strains the heart muscles.

- Smoking–the chances of a heart attack are about 70 per cent higher if one smokes cigarettes.
- Lack of exercise–the physically inactive have a higher risk ratio; everyone should start a program of regular exercise and physical activity.
- Diabetes and a family history of heart trouble–take notice of your family's medical history.
- Anxiety.

Most heart attacks do not cause instant death. A patient may collapse, have trouble breathing, and gasp for air. The victim has intense chest pain and may or may not lose consciousness. The face is pale and the patient may perspire and vomit.

Obtain medical help immediately and encourage the patient until help arrives. Give the patient plenty of air and provide oxygen if it is available. If the victim carries nitroglycerin tablets, place one beneath the tongue. Loosen tight clothing and help the patient avoid strain.

Cancer[2,3]

Cancer is the second largest fatal disease in America, accounting for 635,000 new cases and 300,000 deaths each year; the number of cancer deaths has increased steadily in the past fifty years. The most recent National Cancer Institute report suggests that 80–90 per cent of cancers are caused by environmental pollutants.

Lung cancer kills more men and breast cancer more women than any other forms of the disease. Other common areas of affliction include the colon and rectum, uterus, larynx, bone, and prostate gland.

Cancer is a growth of abnormal cells that invade other tissues and organs. Over one hundred types of cancer have been classified, but full knowledge concerning causes and cures is not known. The diagnosis of cancer in early stages is difficult because many cancers are not visible. It may appear as an unusual lump on the body or it may require an X ray to reveal. Women, especially, should have a pap test at least once a year to ensure against cervical cancer.

Pain is not an unusually early warning sign. However, possible warning signals include:

- A lump or thickening in the breast or elsewhere.
- Unusual bleeding or discharge.
- A sore that does not heal.
- A change in bowel or bladder function.
- Hoarseness or cough.
- Indigestion or difficulty swallowing.
- Change in a wart or mole.

If any of these symptoms appear, consult a physician immediately. A delay in detection means the disease is bringing one closer to death.

Kidney Disease[4]

More than 8 million Americans suffer from kidney disease and urinary tract infections, accounting for 60,000 deaths annually. Infections are the most common form of kidney disease and when left untreated cause severe kidney damage and failure. Bacterial infections impede the flow of urine along the urinary tract. These obstructions can occur at any age, but can be corrected through early detection. The warning signs include:

- Burning or difficulty during urination.
- Frequent urination, particularly at night.
- Passing blood through the urine.
- Puffy eyes or swollen hands and feet (especially in children).
- Pain in the back just below the ribs (not aggravated by movement).

Diabetes

According to the National Commission on Diabetes, 10 million Americans suffer from diabetes; half of these people do not know they have it. The disease occurs when the body does not have enough insulin to properly metabolize sugar; therefore the person's blood sugar level is elevated. A diabetic may feel tired, have to urinate frequently, and have abnormal hunger or thirst. The chances of being diabetic are increased if:

- One is overweight.
- Over forty.

- Has a family history of diabetes.
- If mother bears large babies.

The best way to discover diabetes is to have a blood test given by a doctor. A thorough checkup every year is the best way to discover any disease and to improve one's chances of maintaining excellent health.

Notes

1. "Why Risk Heart Attack?" a brochure by the American Heart Association, 44 East 23rd St., New York, NY 10010.
2. "Possible Signs of Cancer," a pamphlet by the National Cancer Institute, Old Georgetown Road, Bethesda, MD 20014.
3. "Science and Cancer," by Dr. Michael B. Shimkin, a pamphlet by the National Institute of Health, Superintendent of Documents, U.S.G.P.O., Washington, DC 20402.
4. "Urinary Tract Infections," from the National Kidney Foundation, 116 East 27th St., New York, NY 10016.

Additional Sources

The Complete Home Medical Encyclopedia, by Harold Hyman, M.D., Avon Books, New York, NY, 1969.

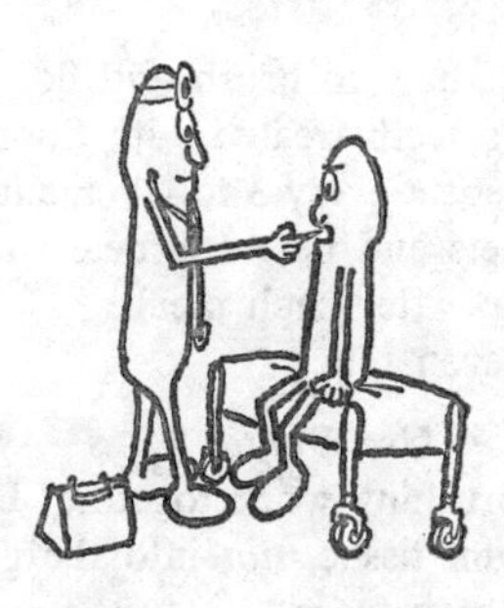

#74 CARE FOR TEETH

Tooth decay is the most prevalent disease in America, afflicting 98 per cent of the population. By age fifteen, the average American child will have eleven decayed teeth and have lost one or more of his/her permanent teeth.

Tooth decay is caused by bacterial action on less exposed

parts of the teeth. One should refrain from eating refined carbohydrates, sweets, and starchy foods which are transformed by bacteria into acids that soften and destroy the enamel. Caries is most likely to occur where the teeth touch each other and on the grooved teeth surfaces.

A cavity cannot heal itself; a dentist must clean, repair, and fill the decayed tooth. Dental care costs Americans about $5 billion a year.

Reduce these costs by taking proper care of teeth. Because the costs of repair far outweigh the costs of extended care, everyone should practice *preventive* rather than *corrective* dentistry. Clean the teeth and gums after each meal by brushing, flossing, and rinsing. Brushing removes plaque from exposed surfaces, and flossing removes food from between the teeth. Brushing with toothpaste will help prevent tooth decay; however, baking soda works just as well and will also neutralize acids that cause bad breath.

Tips for Dental Care[1]

- Visit the dentist twice a year for an examination and correction of defects.
- Ask the dentist how to brush and floss properly.
- Have children's teeth treated with fluoride at three years of age and repeat every 3 to 4 years until age 13.
- Eat fewer sweets and between-meal snacks.
- Brush regularly—after each meal.
- Correct an uneven bite.

Use Dental Floss[2]

- Flossing and brushing work together; both are necessary.
- Floss to the gum tissue, *not* into the gum.
- Clean the tooth surfaces.
- Do not use frayed or soiled floss.
- Rinse thoroughly after flossing to remove loosened food particles and plaque.
- Floss at least every twenty-four hours; make it a habit.

Notes

1. "Healthy Teeth," pamphlet by the Division of Dental Health, Public Health Service, Superintendent of Documents, U.S.G.P.O., Washington, DC 20402.
2. Ibid.

Additional Sources

"Save Your Teeth," Bureau of Health Resources Development, Bethesda, MD 20014.

"Your Guide to Oral Health," a pamphlet by the American Dental Association, 211 East Chicago Ave., Chicago, IL 60611, 1975.

The Tooth Trip: an Oral Experience, by Thomas McGuire DDS, Random House/Bookworks, Berkeley, CA, 1972.

Understanding Dentistry, by Minna Lantner and Gerald Bender DDS, Beacon Press, Boston, MA, 1969.

National Institute of Dental Research
5600 Fishers Lane
Rockville, MD 20852

#75 WATCH WEIGHT

Being overweight or obese (extremely overweight) is a serious health problem in America. Over one-half of the population weighs at least 10 per cent more than ideal for their particular age, sex, and height. One fourth of the adult men and one third of the women in the United States and Canada weigh 20 per cent more than their recommended weight and are considered to be obese. In both sexes, the ages from forty to fifty-nine have the largest number of overweight people.[1]

An estimated 30 to 40 per cent of overweight men consider their weight to be suitable and fewer than half of them have tried to reduce. However, half of all American men will probably die of cardiovascular diseases, and the chances are even worse if the individual has a history of high blood pressure, obesity, high cholesterol, and smoking.

Overeating and lack of exercise are the usual reasons for obesity. The result is that fatty deposits form in body tissues. This condition usually comes from a lifetime of overeating which often begins in childhood.

The obese suffer discomforts from a lack of mobility, even for simple tasks such as tying a shoelace. They often have difficulty buying clothes; they suffer in a society that glorifies the slim and trim. More seriously, they have a decreased life expectancy, being more prone to diabetes, hypertension, and degenerative cardiovascular diseases. The overweight are twice as likely to die from appendicitis and gallstones.

If overweight, realize the seriousness of this condition. Consult a doctor and follow directions for diet and exercise, since reducing plans must be tailored to the individual. Develop habits of eating less, taking smaller helpings, and eating fewer snacks. Carbohydrate intake must be drastically decreased, and one can reduce significantly by eliminating candy, pastry, and sugary desserts from the diet.

Many reducing diets are low in fats. Substituting vegetable oils (safflower, corn, cotton seed, soybean) for animal fats (butter, lard) is a good dietary approach since vegetable oils are lower in saturated fats. Saturated fats are high in calories and tend to raise blood cholesterol levels. High blood cholesterol is a major risk factor in cardiovascular diseases.

Eating breakfast is important in controlling food intake. The calories will be used during morning activity. Remember, calories that are not burned through daily exercise are stored as fat.

Even average-weight Americans in good health can afford to cut down on food intake. Try fasting one day a week, and contribute the savings to organizations feeding the world's starving.

Note

1. *Obesity and Its Management,* by Denis Craddock, Lorryman, Inc., New York, NY 1973.

Additional Sources

"Nutrition, Weight Control and You," prepared by Weight Watchers International, 800 Community Dr., Manhasset, NY 11030.

"Male Myopia About Overweight," by Jean Mayer, the Washington *Post,* July 17, 1975.

The Prudent Diet, by Iva Bennett and Martha Simon, Bantam Books, Inc., New York, NY, 1974.

Food, Nutrition and Diet Therapy, by Marie Knause and Martha Hunscher, W. B. Saunders Co., Philadelphia, PA, 1972.

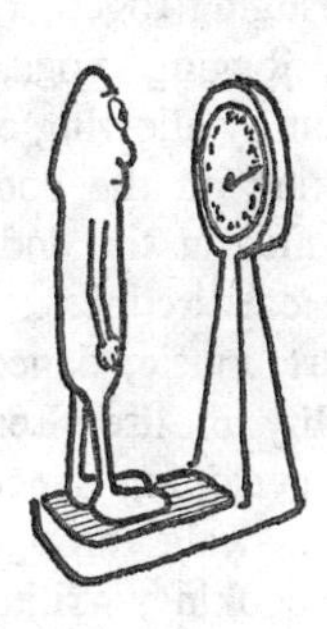

#76 LEARN TO REST AND RELAX

Americans give little value to rest and relaxation. The work ethic reinforces their desire to work at play as strenuously as they work at their jobs. Leisure or after-work activities involve similar stresses. Too many people are unable to unwind because of the tensions that mount during a pressure-filled day. Anxieties, tensions, and irritability are the price the ambitious person pays for success and recognition. Temporary relief is often found in a pill or a drink. Psychological attachment to this solution leads to problems instead of relaxation.

There is nothing wrong with the traditional resting method –taking off shoes, loosening the belt, lying down, and resting. For many people, a few minutes of total rest–at any time of day–will provide ample refreshment to tackle the remaining hours.

However, pleasure and relaxation can also be achieved through the development of self-awareness and the cultivation of a hobby or recreational interests. The most satisfying relaxation will result from opening up oneself to the inner world of the mind.

Physical recreation for pure fun is an excellent way to unwind. Group exercise—volleyball, basketball, interpretive dance, ballet—can provide release from pent-up pressures or nervous energy while bringing together new friends. Individual exercise—swimming, jogging, yoga, and stretching exercises—can bring relaxation by allowing one to forget about job concerns and concentrating on the body, striving to achieve total fitness. Feeling drained at the end of the day is often a mental rather than physical tiredness.

Enjoying aesthetics—art, music, dance, nature—are ways to find peace and tranquility in life. Learn to appreciate the works of others while developing a personal outlet: play an instrument, weave, or just walk through the woods.

Exercise creativity by cooking, writing, sketching, or painting. One may want to relax by reading. Literature provides escape from daily routine, and reading can broaden one's perspectives and thoughts.

Another way to achieve mental rest is through meditating. Meditation frees one from trivial thoughts and outside disturbances and is an excellent way to resolve depression and anxiety.

Dr. Herbert Benson in his book, *The Relaxation Response*, explains how individuals can teach themselves to meditate. From age-old Eastern and Western religions, cults, and secular practices, he extracts the basic components necessary to bring forth the relaxation response:

(1) *A Quiet Environment.*

(2) *A Mental Device.* The repetition of a word or phrase is a way to help break the train of distracting thoughts.

(3) *A Passive Attitude.* Don't worry about how well the technique is performed. Adopt a "let it happen" attitude.

(4) *A Comfortable Position.* The physiological changes elicited are: a decreased rate of breathing, lower heart beat, and reduced blood pressure.

A society that is less tense, pressured, and competitive will develop if we make an effort to unwind and relax. Reduce anxiety and tension, find more enjoyment in life, and improve mental and physical health by learning to rest and relax.

Relaxation Hints

- Lie on back with arms and legs away from the body. Close eyes and consciously begin to relax the muscles in the feet, ankles, and calves. Concentrate only on relaxing each muscle and continue up the body to the thighs, stomach, chest, arms, face, and head. Finally, concentrate only on breathing. Breathe deeply for about fifteen minutes.
- Resolve to spend more time in meditation.
- Get away to a retreat once or twice a year.
- Help others to see the need for relaxation.
- Experiment with different methods of praying.

References

The Relaxation Response, by Herbert Benson, M.D., William Morrow and Co., New York, NY, 1975.

"Valiumania," by Gerald Cant, *The New York Times Magazine,* February 1, 1976.

Prayer and Meditation, by F. C. Happold, Penguin Books, Baltimore, MD, 1971.

Additional Sources

Resistance and Contemplation, by James W. Douglas, Dell Publishers, New York, NY, 1973.

The Climate of Monastic Prayer, by Thomas Merton, Cistercian Publishers, Kalamazoo, MI, 1975.

Seeds of Contemplation, by Thomas Merton, New Directions Books, Norwalk, CT, 1949.

Stress Without Distress, by H. Selye, J. B. Lippincott, New York, NY, 1974.

The Buddha's Way, by H. Saddhatissa, Allen and Unwin, London, Eng., 1971.

On the Psychology of Meditation, by C. Naranjo and R. E. Orstein, Viking Press, New York, NY, 1971.

The Science of Being and Art of Living, by Maharishi Mahesh Yogi, International SRM Publications, London, Eng., 1966.

Christian Zen, by W. Johnston, Harper & Row, New York, NY, 1971.

Yoga, Immortality and Freedom, by Mircea Eliade, Princeton University Press, Princeton, NJ, 1973.

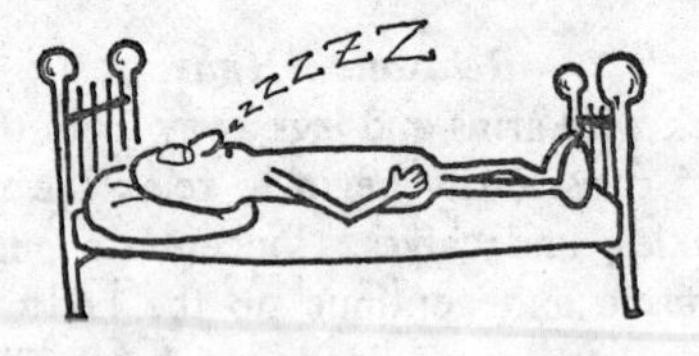

IX TRANSPORTATION

The American Dream Machine–the automobile–once meant convenience, easy transportation and pleasurable outings. But that dream is rapidly becoming a nightmare created by the very vehicle that promised and gave us so much. The growing army of automobiles has diminished the quality of our lives by polluting our air and our environment. The advantages are being eroded by the disadvantages. Our air is not fit to breathe, traffic is jammed bumper to bumper, noise assaults us from the roadways. And now, gasoline shortages compound our problems.[1]

INTRODUCTION

Moving people and freight accounts for about a quarter of America's national energy budget. We move about more than any other people, and we use low-efficiency means to do much of our traveling. The total energy demand for automobiles alone, including petroleum refining and retailing, vehicle manufacture, repairs, tires, insurance, and highway construction, accounts for over 21 per cent of the U. S. energy budget.[2] The ubiquitous automobile has not only depleted our oil reserves but has also drained our mineral supplies. In addition, excessive amounts of energy are utilized in street maintenance, traffic control, motor vehicle code enforcement, parking facilities, and junk-car disposal.

Perhaps even more serious are the costs to the environment. Air pollution alone causes an estimated $16 billion a year in damage to human health, crops, livestock, buildings, and property values. Noise pollution, traffic congestion, and highway construction have also contributed to the ecological damage of our country.

As a people mover, the automobile is highly inefficient. The urban automobile, as it is currently used, consumes about twice as much energy per passenger/mile as does a bus.[3] Mass-transit systems utilizing bus and rail offer the most efficient means of passenger travel. The bicycle still ranks first in operating efficiency (in terms of energy used in moving a certain distance as a function of body weight). For short trips to school, work, church, or shopping, the bike is the best way to go. Apart from the benefits of exercise, the bicycle produces no pollutants and consumes no gas. The materials used in a bicycle amount to 25 pounds of steel, three pounds of rubber, and a few bits of plastic, leather, and paint. Compare this with the average car which requires 3,500 pounds of steel and a myriad of accessories.

The bicycle and mass transit will go a long way in stretching our energy resources, but those who drive must also conserve by maintaining an efficient automobile. A few effective ways to reduce gasoline consumption are: 1) share the car; 2) avoid long-distance travel; 3) purchase only a small, economic car; 4) use the proper gasoline; 5) drive conservatively; and 6) maintain and service the car periodically.

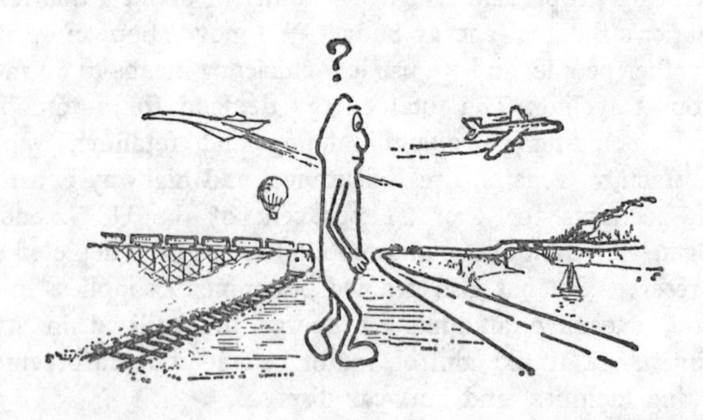

Notes

1. "Clean Air and Your Car," Environmental Protection Agency Report, Washington, DC 20460, March 1974.

2. *Lifestyle Index,* by Albert Fritsch and Barry Castleman, CSPI Publications, 1757 S St., NW, Washington, DC 20009, 1974, p. 33.
3. *Science,* Vol. 189, July 11, 1975, p. 98.

#77 KNOW BICYCLE BENEFITS AND SERVICING

America is experiencing a boom in bicycle sales. According to the Bicycle Manufacturers' Association of America, there are 75–80 million bicycles in the United States today. The rapid rise in bike use during the '70s has broadened the bicycle's role, from a recreational vehicle to a practical transportation alternative.

As a people mover, the auto is highly inefficient. Statistics show that over one half of all auto trips cover distances less than five miles.[1] In an urban area, these short distances could easily be traveled by bicycle in 30 minutes, which in most cities is comparable or better than the time it takes to cover the same distance by car. In terms of energy use, the bicycle is the most efficient means of travel on earth.[2] The bicylist traveling at 10 mph is reaching the equivalent of 1,000 passenger-miles-per-gallon of petroleum fuel–forty times more efficient than motorized transports.[3]

Two obvious results of increased bike use are reduced gasoline consumption and reduced auto emissions. Theoretically, if all urban trips shorter than four miles were converted to bicycle trips, roughly 8 per cent of the vehicle-miles traveled would be curtailed, which would result in a 7.6 per cent saving in the energy consumed by transportation.[4] In urban areas, where short bicycle trips are practical, most air pollution is caused by auto emissions. Therefore, increasing the number of commuter cyclists could have a significant impact on reducing air pollution. According to an Environmental Protection Agency study, emission reductions could range from 2 to 10 per cent depending on bicycle use.[5]

The bike is a relatively inexpensive mode of travel. Initial bicycle cost is low compared to the auto, and maintenance costs are trivial considering the exorbitant costs of auto repair. The differences in resource utilization and economic costs between bikeway construction and highway construction are also great.

Other less-quantitative benefits of cycling include reduced traffic and parking congestion, less noise pollution, better-quality air, greater mobility in business districts, reductions in road building, and healthy exercise. Especially during the rush hours, traffic congestion could be significantly reduced if bicycle trips replaced motor vehicle trips. Twenty bikes can fit into a parking space used by one car. If bikes were widely used, on-the-street parking and the need for more roads could be eliminated in most urban areas. Noise pollution from motorized vehicles can be eliminated by the silent bicycle, and air quality could be significantly improved. Another positive inducement for biking is its tremendous versatility. The bike provides door-to-door convenience. Medical literature contains numerous accounts hailing the physical and psychological benefits of biking. Cycling helps strengthen the heart, lower blood pressure, increase circulation, and control weight.

How to Keep a Bike

In 1974 an estimated half million bikes were stolen. Bicycle registration and licensing is a must, but additional insurance against theft is the right kind of chain and lock. Most enthusiasts recommend a case-hardened ⅜″ chain with a strong, secure padlock. To prevent marring, encase the chain in an old innertube. When locking a bike, be sure to lock the frame and both wheels to an immovable object. A good habit is to lock the bike where pedestrian and auto traffic are heavy. Never leave a bike unattended at night. Be sure to register the bike by serial number, make, and model, so if the bike is stolen the chances of recovery will be greater. Also, list the bike by color, model, and serial number on the personal property floater of the homeowner's or tenant's insurance policy.

How to Buy a Bicycle

Based on need, intended use, and cost, there are many things to consider when buying a bike. One of many books dealing with this subject is Peter Braddock's *How to Choose and Use a Bicycle.* Any bookstore will have this and many other books dealing with bicycle purchasing.

How to Maintain a Bike

Bikes do not need much attention, but they should never be neglected. Clean the chain, axles, and other moving parts periodically with an old toothbrush or cloth soaked in kerosene and then lubricate with a light oil. Keep tires inflated to the proper pressure that is stamped on the tire. Always check the brakes. Make sure they fit properly and are not worn. Spokes should be tightened so that they "ping." Most minor adjustments can be made with a screwdriver, pliers, adjustable wrench, and repair manual. Two good repair books to start with are, *Anybody's Bike Book,* by Tom Cuthbertson, and *Bicycle Repair* by T. C. Kleeburg. If planning a long trip, make an equipment checklist similar to the one below:

________small adjustable wrench, pliers, and screwdriver

________tire pump and gauge

________bicycle lock

________tire repair kit

________tire iron (clinchers)

________spare tire tube

________extra spokes

________extra brake pads

________rear brake cable

________gear cable

________chain rivet tool

________spoke wrench

Notes

1. "Clean Air and Your Car," E.P.A. Report, Washington, DC, 1974, p. 5.
2. *Bicycle Transportation,* by Nina Dougherty and William Lawrence, E.P.A. Report, Washington, DC, 1974, p. 9.
3. Ibid., p. 9.

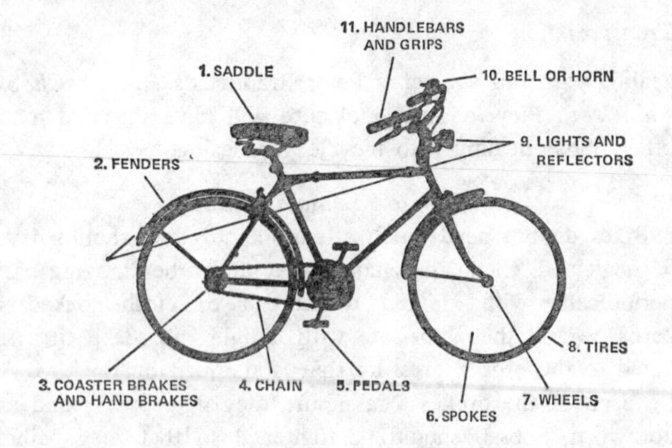

Bicycle Safety Checks

1. Saddle: Adjust frequently for comfort and growth. Tighten saddle and seat post nuts securely, leaving at least 2½ inches of seat post down in the frame.

2. Fenders: Be sure they are undamaged and securely fastened.

3. Coaster Brakes and Hand Brakes: Check before your first ride. Must brake evenly every time, no slippage. The purchaser should periodically inspect and maintain brakes. The coaster brake arm must be securely fastened to the frame.

4. Chain: Chain should be checked frequently for damage and stretch, and be readjusted if necessary. It should be lubricated frequently with light oil.

5. Pedals: This bicycle is equipped with reflectorized pedals for added safety in night riding. It is imperative that the shoulder of the pedal axle be securely tightened against the crank arm. If pedals become worn or damaged, replace them with reflectorized pedals.

6. Spokes: Replace broken ones promptly. Keep them tight.

7. Wheels: Should rotate smoothly without wobbling from side to side. If necessary, they should be realigned. Axle nuts should be kept tight.

8. Tires: Should be inspected frequently for wear and leaks. Remove imbedded stones, nails, glass, cinders, etc. Keep inflated to the correct pressure which is stamped on the sidewall of the tire.

9. Lights and Reflectors: This bicycle is equipped with reflectorized pedals and other reflective materials. If these materials are damaged or lost, replace them immediately for your own safety. Lights and reflectors should be visible at dusk and at night; headlights from 500

4. "Energy Use for Bicycling," by Eric Hirst, A.E.C. Report, Washington, DC, 1974.
5. Dougherty and Lawrence, op. cit.

Additional Sources

Popular Mechanics Book of Bikes and Bicycling, by Dick Teresi, Hearst Corp., New York, NY, 1975.

Glenn's Complete Bicycle Manual, by Clarence Coles and Harold Glenn, Crown Publisher, New York, NY, 1973.

Richard's Bicycle Book, by Richard Ballantine, Ballantine Books, New York, NY, 1972.

The New Complete Book of Bicycling, by Eugene A. Sloane, Simon and Schuster, New York, NY, 1974.

Anybody's Bike Book, by Tom Cuthbertson, Ten Speed Press, Berkeley, CA, 1971.

Bicycle Repair, by Irene Kleebert, Concise Guide Services, Watts Franklin, Inc., New York, NY.

---

feet, rear reflector from 300 feet. Be sure reflectors are state approved.

10. Bell or Horn: Be sure it works properly, loud and clear.

11. Handlebars and Grips: Handlebar should be adjusted frequently for comfort and growth. Keep at least 2½ inches of handlebar stem down in the frame, then tighten it securely. Handlebar grips should fit snugly, and worn ones replaced.

Have your bicycle inspected twice a year by a competent person. (Reproduced with the permission of Bicycle Manufacturers' of America.)

#78 BIKE SAFELY

The number of accidents, injuries, and deaths associated with bicycle riding has increased dramatically in the last few years. The National Safety Council estimated that 1,100 cyclists were killed in 1973. In addition, one million bicycle injuries requiring medical treatment occurred. The major problem is not with the bicycle itself. Many times accidents are the fault of the rider running through a light or a motorist who isn't aware of the cyclist. Another factor that compounds the problem is a traffic environment in which a bicycle cannot hope to compete. Shoulderless roads, narrow streets, a lack of bikeways, and congested traffic greatly impede bike travel.

The majority of cycling accidents involve children, which strongly suggests the need for bicycle-safety training. Bicycle safety should begin in the elementary school and be taught by a professional who knows how to effectively communicate with the child. A helpful guide follows:

SUGGESTED SAFE BIKE DRIVING RULES
(courtesy of Bicycle Manufacturers' of America)

1. *Obey all traffic regulations, signs, signals and markings.*
   Bicycles should be driven as safely as any road vehicle, and they are subject to the same rules of vehicular traffic. A good "rule of thumb" is to avoid congested streets and use bikeways, lanes or paths where possible.
2. *Observe all local ordinances pertaining to bicycles.*
   Registration and licensing, inspections, driving on sidewalks, etc., may be covered by local laws. It is a personal responsibility to know them and abide by them.

3. *Keep right: drive with traffic, not against it. Drive single file.*

   Keep as close to the curb as practical. Most states require bikers to drive single file. When driving two abreast, a minor swerve could force a cyclist into traffic.

4. *Watch out for drain grates, soft shoulders and other road-surface hazards.*

   Be careful of loose sand or gravel, particularly at corners, and watch out for pot holes.

5. *Watch out for car doors opening, or for cars pulling into traffic.*

6. *Don't carry passengers or packages that interfere with vision or control.*

   A good rule is "one person, one bike," unless it's a tandem. Use baskets or luggage carriers for packages.

7. *Never hitch a ride on a truck or other vehicle.*

8. *Be extremely careful at intersections, especially when making a left turn.*

   Most accidents happen at intersections. If traffic is heavy, get off and walk the bike with pedestrian traffic.

9. *Use hand signals to indicate turning or stopping.*

   Give the appropriate hand signals for turning left or right, or for stopping.

10. *Provide adequate protection at night with the required red reflectors and lights.*

    Again, state laws vary. Most require a headlight, taillight or red rear reflectors for night cycling. Others require reflective pedals, additional side reflectors or other reflective material. If riding at night, wear light-colored or reflective clothing.

11. *Drive a safe bike. Have it inspected to ensure good mechanical condition.*

    Make sure the bike fits the cyclist. Make sure that brakes, pedals, lights, reflectors, shifting mechanisms, sounding devices, tires, spokes, saddle, handlebars, and nuts and bolts are checked regularly.

12. *Drive defensively; watch out for the other guy.* Observe the car in front and the one in front of it. Leave room and time to take defensive action.

An often overlooked safety hazard is the dog. Perhaps the best defense is to dismount and walk the bike, keeping the frame between oneself and the animal. When out of the dog's home territory, he probably won't bother the biker.

Since most bicycle fatalities are the result of head injuries, you should wear a *hard* helmet. Brain damage is usually irreversible.

BIKEWAYS

Bikeways separate the cyclist from motor-vehicle traffic and provide a great deal of safety for the bike rider. There are three types of bikeways:

- *Class 1*–A completely separate and exclusive bike path.
- *Class 2*–A lane on a street or road reserved for bicycle use, marked by a painted line, rubber bumpers, concrete curbing, or some other device.
- *Class 3*–A shared route designated only by signs. This bikeway is the most common today.

Pressure local officials to establish bikeway networks and to apply for Federal Highway grants or local public works funds to develop bikeways. Join a bike club and help lobbying efforts concerned with improving the plight of the cyclist.

A great untapped resource for bicyclists is the extensive network of transportation routes that crisscross every part of the nation. These old transportation routes are now abandoned: railroads, canals, aqueducts, and trolleys. They can easily be converted into biking trails, providing transportation and recreation for the community. For a detailed account of the legal and financial steps involved in converting these abandoned routes into bikeways, read "From Rails to Trails," from the Citizens Advisory Committee on Environmental Quality, 1700 Pennsylvania Ave., NW, Washington, DC 20006.

Citizens should also demand more parking facilities for bicycles. Bikes can be integrated into a multifaceted transportation scheme if they can be safely parked at commuter terminals. Information on bikeway routes and tour information is available from:

American Youth Hostels
20 West 17th St.
New York, NY 10011

League of American Wheelmen
3582 Sunny View Ave., NE
Salem, OR 97303

Additional Sources

North American Bicycle Atlas, from American Youth Hostels, 20 West 17th St., New York, NY 10011.

Bike Tripping, from Ten Speed Press, Box 4310, Berkeley, CA 94704.

The American Biking Atlas and Touring Guide, by Sue Browder, Workman Publishing Co., New York, NY.

#79 ENCOURAGE MASS TRANSIT

Comprehensive mass-transit systems that use various modes of transportation (bus, rail, subway, bicycle, and car pool: excluding air travel) can effectively compete with auto travel, provided they are carefully planned and operated. Mass transit can alleviate much of the congestion, air pollution, noise pollution, land development, and energy wastefulness that plagues our society. Canada's second largest city, Toronto, is a good example of how mass transit can be successful. Utilizing a co-ordinated system of rail, subway, and bus travel, the city has provided its residents with high-quality inner-city travel at a low cost ($.25 in 1974).[1]

Airplanes compete with the auto as the most energy wasteful means of transportation, requiring two to four times as much energy per passenger mile as a train. The problem is compounded because most planes carry partial loads, which increases fuel wastage. However, airline load factors are rising dramatically, thus lowering energy per user.

The private auto is just as wasteful. In 1973 almost three fourths of all the gasoline used in this country was consumed by the automobile.[2] From an energy standpoint, the waste inherent in using this much fuel is indefensible, especially in urban areas where buses are nearly three times more efficient per passenger mile than the private auto. A shift from present 97 per cent auto–3 per cent mass transit, to 75 per cent auto–5 per cent bike–20 per cent mass transit, even after converting to fuel-economy autos, would save 0.5 per cent of current total U. S. energy consumption.[3] The auto is particularly energy-intensive because of a low-seating capacity in relation to its weight. On the other hand, a bus can carry a large number of passengers, and is especially efficient during rush hours.

The Demand-Responsive Transportation System or Doorstep Transit is the newest mass transit innovation that promises to overcome the traditional problem of route inflexibility.[4] The broad dispersal of urban and suburban residences, shopping centers, and job locations prompted the idea that instead of making people fit an arbitrary schedule and route, the transit system should suit individual needs. Although these systems bear a variety of names, such as Call-A-Ride, Dial-A-Bus, Telebus, or Dial-A-Ride, they are similar to each other in operation. Riders call their travel requests into a control center, where trips of similar origin and destination are grouped together. With several people sharing the vehicle, usually a van, the cost per trip is relatively inexpensive. The vehicles do not have fixed routes or schedules.

This kind of system combines the convenience of the automobile with the energy efficiency and low cost of mass transit. Cheaper than rail systems, Doorstep Transit is flexible enough to adapt to population shifts. Most importantly, it saves fuel and reduces congestion and pollution. Doorstep Transit also provides easy transportation for the aged, disabled, young, and poor who do not own cars and are not served by mass transit.

IDEAS FOR IMPROVING MASS TRANSIT

Buses and Trains:

- Improve ventilation and temperature control in buses and trains.
- Lower fares.
- Provide more extensive routing.
- Provide express bus lanes.
- Run free minibuses in the downtown shopping district.
- Ban all traffic except buses and bicycles from downtown areas.
- Push for less highway spending, giving more money to bus and rail systems.

Auto Disincentives:

- Charge a commuter tax on cars.
- Eliminate free parking and divert revenue into mass-transit systems and bike facilities.
- Charge graduated highway and bridge tolls depending on the number of passengers.
- Inspect air-pollution devices and levy a fine if they do not work properly, channeling revenue into mass-transit systems.
- Impose an excise tax on large cars.

• Develop schemes and strategies to make the automobile rider subsidize the mass-transit rider, not vice versa as current practices dictate.

Fighting Superhighways

In most American cities, new highway construction remains a major factor in disrupting the living environment. Last year, almost one out of every six GNP dollars was connected with highway-related activities. In a normal year highway construction displaces 50,000 people. Highways are also an indirect source of most urban air and noise pollution.

Effective community organizing is what makes or breaks a highway struggle. An excellent account of how community pressure was brought to bear on a city hall is *Rites of Way: The Politics of Transportation in Boston and the U. S. City,* by Alan Lupo, Frank Colcord, and Edmund Fowler. The bible for highway fighters is *Superhighway; Superhoax* by Helen Leavitt. The Center for Science in the Public Interest has also published a report entitled *Highways and Air Pollution,* which may be helpful to the highway activist. Organizations dealing specifically in highway problems are:

Highway Action Coalition
1346 Connecticut Ave., NW
Washington, DC 20036

Clean Air Campaign
Marcy Benstock
11 West 42nd St.
New York, NY 10036

Notes

1. "Making Mass Transit Work," *Business Week,* February 16, 1974, p. 74.
2. "Clean Air and Your Car," E.P.A. Report, Washington, DC 20460, March 1974.
3. *The Contrasumers: A Citizens Guide to Resource Conservation,* Albert Fritsch, Praeger Publishers, New York, NY, 1974, p. 129.
4. "A Ride for Everyone," by John Newman, *Environment,* June 1974, pp. 11–18.

Additional Sources

"Door Step Transit," by Daniel Roos, *Environment,* June 1974, pp. 19–28.
Transportation Special I, Environmental Action, December 6, 1975.
Transportation Special II, Environmental Action, December 13, 1975.
Potential for Energy Conservation in the United States: 1974–1978,

Transportation, A Report of the National Petroleum Council, September 10, 1974.

Superhighway; Superhoax, by Helen Leavitt, Ballantine Books, New York, NY, 1970.

Highways and Air Pollution, by James Sullivan, CSPI Publications, 1757 S St., NW, Washington, DC 20009.

#80 SHARE THE CAR

Every day 58 million American workers rely on the automobile for transportation to and from work. Of these, 40 million drive alone, consuming 290 million gallons of gasoline each week. Since commuting to work is the largest form of auto use, it is a good place to start conserving energy. Car pooling is one method to achieve this aim. A few benefits of car pooling include:

- Saving energy–An increase in occupancy from the current 1.6 to 2 persons per car would result in an annual national fuel savings of almost 5 billion gallons of gasoline.
- Improving air quality–Vehicle exhaust accounts for 85–95 per cent of the pollutants in the air. High levels of pollutants are especially present during morning and evening commuter traffic. Fewer cars on the road would lead to lower pollution levels and cleaner air.
- Reducing traffic congestion during rush hour–If occu-

pancy can be increased during commuter trips to 2 persons per car, then 20 per cent of the automobiles could be taken off the road. Reduced traffic on roads would shorten the length of the rush hour and the amount of time it takes to get to and from work.

- Providing alternative service for non-drivers—Car pooling affords transportation to non-drivers where mass transit is not available.
- Financial savings—Car pooling can save $500 a year or more; one may even be able to give up a second car. If one spends $2 a day on gasoline, and parking costs another $2, that's $4 a day or $80 a month commuting tab. If one shares the bill with another rider, the costs will plummet to $40 a month. With 3 others, it costs $20, and with four other persons the cost is a mere $16 per month per person. Savings increase as the commuting distance grows and the price of gas increases.
- Reduced anxiety—On non-driving days a car pooler can use the extra time to nap, read, or converse. There is no need to worry about fighting traffic.
- Time saving—Many communities across the country have developed special commuter lanes where cars with four or more passengers ride in an exclusive express lane. This reduces the time getting into the business district. Once there, a car pooler may find that many employers offer special parking privileges for car poolers.
- Benefits to employers—Car pooling decreases the number of parking spaces required by employees, thus reducing the operating costs of a firm. People who car pool tend to be on time for work because they are reluctant to keep others waiting and are afraid of missing a ride if late. Many employers testify that employees who participate in car pools have better attendance records.

Starting a car pool is much easier than one might think. Many times it's as easy as talking to two or three neighbors who go to work at approximately the same time and in the same vicinity.

If a good, compatible group is assembled, discuss the advantages of leasing or buying a community car or van. This could eliminate the necessity for a second family car.

Another approach is to get employers involved. Point out the benefits of car pooling and ask if the company would help coordinate a car-pooling effort. This type of approach has been very successful in California.

In Minneapolis, the 3M Company bought 76 vans and recruited employees to drive them. The driver paid nothing for the ride and was able to use the van at night and on weekends. He/she also could keep the daily fees of any riders over ten; the first ten commuters paid enough to operate the van. The Federal Energy Administration estimated that one van saved approximately 3,738 gallons of gas in a year.

Additional Sources

Commuter Computer, Suite 610, 3440 Wilshire Blvd., Los Angeles, CA 90010.

Double Up, America, report by the Federal Highway Administration, Superintendent of Documents, U.S.G.P.O., Washington, DC 20402, April 1975.

How to Pool It, report by the Federal Highway Administration, Superintendent of Documents, U.S.G.P.O., Washington, DC 20402, May 1975.

#81 AVOID UNNECESSARY AUTO TRAVEL

The average American car is driven about 13,000 miles each year. If we were budgeted by "mileage rationing" (each vehicle owner allowed a predetermined mileage allowance for one year with excess mileage subject to a federally imposed surtax), we might be more conscious of our trips.[1] Drivers often take trips that need not be made. Pleasure rides, visits to friends and relatives, recreational trips, and vacations consume 382 million gallons of gasoline each week.[2]

The pleasure ride is pure extravagance in an age of air pollution and fuel shortages. If someone seeks pleasure, there are numerous leisure options that are unmotorized (see Section VII). Take a pleasure stroll or ride a bike.

Visiting the sick, elderly, relatives or friends is an act of kindness and should never be discouraged. When the visit is for an informal chat or business, a telephone call or a letter is far less costly in energy resources. Decide in advance how important the visit is, and try to combine it with shopping and business trips, or travels to and from work.

Vacations are large auto energy users, and too often we drive to distant locations to vacation with crowds of other people. Vacation time is wasted by driving great distances, and the tension and strain can be greater than that of the daily work routine. When planning a summer vacation, investigate the recreational possibilities within a limited geographical area, say a 200-mile radius. Seek spots that offer quiet, fresh air, and a change of scenery. Most Americans can find these vacation conditions close to home.

Another category of potentially wasteful travel centers around short trips and errands such as shopping, civic, educational, and religious activities, visits to the doctor and dentist, and transporting children to and from school activities and recreation sites. These short trips consume over 225 million gallons of gasoline each week in this country.

Frequent shopping trips cost us money and energy. An average family goes to a store, laundromat, hairdresser, or similar place at least once a day. If the store is more than two blocks away, the trip is probably made by car. Walk, if the destination is a mile or less; also, shop on the way home from school or work. Concentrate shopping into a single weekly event at one location if possible. Remember that short trips are hard on gas mileage and frequently result in auto accidents—the majority of which are within a mile of home.

They also contribute to neighborhood pollution and congestion.

Car pooling, as mentioned in #80, can be an effective energy saver for transporting children to school and for attending evening classes, church gatherings, and social meetings. For many of these, one may not need to use a car. Teach others to walk and bike. Provide neighborhood kids with employment and responsibility by asking them to run errands. Auto savings would be converted to human service and give young people an opportunity to be useful.

AUTO MILEAGE BUDGET

Original Estimate	Current Year	Next Year	
			Work
			Store
			School
			Church
			Pleasure-Riding
			Vacation
			Recreation

Keep a record of mileage used for the various categories. The best way to conserve is to see exactly where mileage is spent. Before keeping tabs, write an estimate. The totals might be surprising, and provide an incentive to save.

Notes

1. Suggestion made by Charles E. Donnelly, Old West Trail Foundation, Rapid City, SD, in a letter to F. Zarb, Federal Energy Administration, January 29, 1975.
2. "Gas Watchers' Guide," American Automobile Association, Falls Church, VA 22042.

#82 CAR-BUYING HINTS

Since we use more energy for cars than for any other single purpose, buying an efficient car (provided it is necessary) can be an important conservation effort. Buy only as much car as is needed. A few questions one should ask are:

- Is the car for long-distance commuting or just for short trips to the store?
- Will it be driven primarily on open highways or on congested city streets?
- On an average, how many people will be riding in the car?

Vehicle weight—Vehicle weight is the single most important item affecting fuel economy. A 5,000-pound car takes twice as much energy to run as a 2,500-pound car and gets half as much mileage. The Environmental Protection Agency estimates that if we reduced the average weight of our cars from the present 3,500 pounds to 2,500 pounds, America would save 2.1 million barrels of oil a day, about a third of current vehicle fuel consumption.[1]

Tires—Tests have proven that radial ply tires significantly reduce roadway friction. They cost a little more initially, but wear longer and provide up to 10 per cent better mileage than conventional tires.

Engine Size—The most economical engine is the 4-cylinder, which is adequate to handle normal auto travel.

Air Conditioning–An air conditioner (used in more than 60 per cent of new cars) can reduce gas mileage by 9 to 20 per cent, depending on driving and weather conditions.[2]
Automatic Transmission–Depending on the owner's driving ability, a manual transmission can save 2 to 15 per cent of the fuel needed to run an automatic transmission car.[3]
Extras–Optional extras add to gasoline consumption. Frills such as power steering and power brakes should not be considered lightly, because their combined effects can devour significant amounts of gas. Power windows are an unnecessary extravagance.

Watch out for gasoline guzzlers when buying a car. It could mean a thousand or more dollars over the lifetime of the car. Ask motorists who drive a similar car how many miles they get per gallon.

According to E.P.A. test findings, 1976 autos averaged 12.8 per cent better fuel economy than 1975 cars. New cars average 17.6 miles per gallon. Cars that ranked highest in terms of miles/gallon are:

- 33 mpg–Chevrolet Chevette (M); Datsun B–210 (M); Subaru 83CID (M).
- 32 mpg–Chevrolet Chevette 85CID (M); Renault 5 (M) 79CID.
- 30 mpg–Peugeot 504 Diesel (M) 129CID; Peugeot 504 Diesel (M) Wagon 129CID.
- 29 mpg–Audi Fox 97CID (M); Audi Fox Station Wagon 97CID (M); Austin Morris MG Midget 91CID (M); Chevrolet Chevette Automatic 98CID; Datsun B–210 85CID (A); Subaru Wagon 83CID (M); Triumph Spitfire 91CID (M); Volkswagen Dasher 97CID (M); Volkswagen Dasher Wagon 97CID (M).
- 28 mpg–Audi Fox 97CID (A); Audi Fox Station Wagon 97CID (A); Ford Pinto 140CID (M); Subaru 97CID (A); Subaru Wagon 97CID (A); Toyota Corolla 97CID (M); Toyota Corolla Wagon 97CID (M); Volkswagen Dasher Wagon 97CID (A).

(M)–Manual
(A)–Automatic
CID–Cubic-inch displacement

These ratings are a composite of two tests that incorporate city and highway mileage figures. They are based on Federal Highway Administration estimates that the average consumer does 55 per cent of his/her driving under city conditions and 45 per cent on the highway.[4]

Notes

1. "Crash Diet for Overweight Cars," *Energy Reporter,* FEA Citizen Newsletter, Washington, DC 20461, October 1975, p. 4.
2. "Clean Air and Your Car," E.P.A. Report, Superintendent of Documents, U.S.G.P.O., Washington, DC 20402, March 1974.
3. "Motorist's Guide to Saving Gas," Publication of the Office of Energy Conservation, Toronto, Can., 1974.
4. *Gas Mileage Guide for New Car Buyers, 1976,* Federal Energy Administration, Pueblo, CO 81009.

Additional Sources

How to Buy a Used Car Without Getting Gypped, by Peter Hahn, Harper & Row, New York, NY, 1975.

#83 CHOOSE PROPER GAS

A car performs more efficiently when it uses a gasoline with the proper octane rating. Buy the lowest octane fuel that will keep the engine from knocking. Knocking decreases power and fuel economy, and can damage engine parts. Experts estimate that 40 per cent of all drivers purchase fuel with an octane rating that is too high.[1]

Don't believe the myth that premium means better or that an occasional tankful of high test will blow the carbon out of the engine. Owner's manuals recommend specific octane grades, but these refer to the average car coming off the assembly line. An individual car may be above or below average. Furthermore, a car's octane requirement may change with age and lead build-up in the engine.

Experiment through buying a low-octane gasoline. If no knocking occurs, buy a lower grade. Try mixing two grades: ½ tank of one and ½ of another, or ¼ of high octane and ¾ of low.

Octane numbers vary according to different rating systems. The ones listed in older operator's manuals are *RON* numbers, which are higher than those posted on pumps today.

	Manual Octane Rating (RON)	*Federal Pump Octane Rating*	*Application*[2]
ECONOMY AND UNLEADED	91–92	87–88	For many 1971 and newer models.
REGULAR	93–95	89–91	For most 1970 and prior models designed to operate on regular gasoline.
MIDPREMIUM	96–98	92–94	Intermediate grade.
PREMIUM	99–101	95–97	For high-compression-ratio engines that run on premium gasoline.

When choosing a gasoline, do not let brand loyalty make the decision. Traditional come-ons and familiar signs do not mean better fuel. There is no tiger in anyone's tank. Unknown brands pump the same gasoline as the major brand down the street. Strict company control of fuel from oil well to station pump is the exception rather than the rule. A tankful of gasoline will vary more region to region than by brand. Buy according to price—not brand.

Choose a responsible dealer with a heavy volume of gasoline sales. Fuel that remains in the station pumps can accumulate dirt, water, and harmful polymers that affect auto performance. Make sure there are filters on the dealer's pumps. Find out how dependable the station attendants are. Prompt and courteous service can indicate proper fuel tank maintenance.[3]

Buy unleaded gas whenever possible, especially for new model cars. Leaded gas destroys the catalytic converter (the emission-control device). According to the Environmental Protection Agency, lead emissions from the tail pipe account for 90 per cent of the lead emitted into the atmosphere. This can be breathed or can enter the human body in water, food, contaminated dust, and dirt. Many health officials are concerned that children playing near heavy traffic are exposed to high levels of lead pollution.

Do not let the service station attendant fill the tank after the pump shuts off. From 10 to 40 million gallons of gasoline are spilled each year through this practice, which is enough gas to run 55,000 cars for one year.[4]

Notes

1. *The Contrasumers: A Citizen's Guide to Resource Conservation,* by Albert J. Fritsch, Praeger Publishers, New York, NY, 1974.
2. "A Consumer's Guide to Octane Ratings," Connecticut Energy Agency, Hartford, CT, 1975.
3. Fritsch, op. cit.
4. *Energy Reporter,* F.E.A. Citizen Newsletter, Washington, DC 20461, January 1976, p. 5.

Additional Sources

"Investigation of Passenger Car Refueling Losses," Government Publication #PB21592, National Technical Information Service, Port Royal Road, Springfield, VA 22161 ($5.75).
"Guide to Octane Rating," Reprint from *Consumer News,* Dept. of Transportation, Office of Consumer Affairs, Washington, DC.
"Gasoline: More Miles Per Gallon," Department of Transportation, Office of Consumer Affairs, Washington, DC.

#84 DRIVE CONSERVATIVELY

Driving speeds and habits have a tremendous impact on gasoline efficiency. Most interstate highways now have limits of 50–55 mph, but it is the individual's responsibility to obey these laws. These limits reflect a concerted effort by local, state, and federal governments to improve the safety of our highways and maximize fuel conservation. The average car driven between 75 and 80 mph will consume almost twice as much fuel per mile as the same car driven at 50 mph.

Studies conducted by one American Automobile Association club showed that gasoline efficiency could be increased by as much as 44 per cent if driving habits were improved.[1]

- Don't idle–When starting the car, don't leave the motor idling to warm up. Start driving when the engine is running smoothly (1 minute) and drive the first mile at reduced speed to allow the engine to stabilize. Racing the motor wastes gasoline and damages the engine. Idling more than one minute will waste more gas than it takes to restart the engine.
- Accelerate slowly–Smooth footwork is crucial to good mileage. Avoid jack-rabbit starts and stops by building up speed slowly and coasting to a stop. Pretend there's an egg between the foot and the gas pedal. This simple practice can save up to 25 per cent on gas bills.
- Shifting–Proper shifting can save a mile per gallon.

- Avoid heavy traffic–Stop-and-go traffic wastes more gas than driving at excessive speeds.
- Unload trunk–Hauling excess baggage puts an added burden on the engine. If the trunk contains tires (other than spare), heavy tools, sporting equipment, and excess weight, clear it out.
- Travel at moderate speeds–Driving at high speeds requires more energy to overcome increased air resistance.
- Don't overfill the gas tank.

Note

1. "Gas Watcher's Guide," American Automobile Association, Falls Church, VA, 1975.

Additional Sources

Potential for Energy Conservation in the United States: 1974–1978 –Transportation, A Report of the National Petroleum Council, 1625 K St., NW, Washington, DC, September 10, 1974.

"Strict Auto Licensing May Curb Energy Use," ECP Report, Environmental Law Institute, 1346 Connecticut Ave., NW, Washington, DC 20036, December 1975.

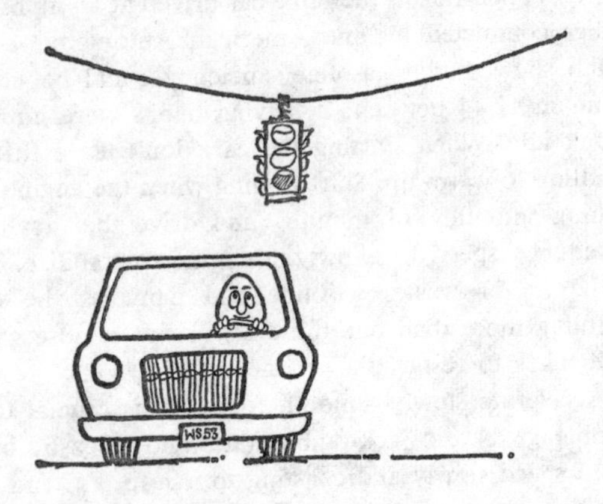

#85 SEEK PROPER MAINTENANCE AND SERVICE

If car travel is necessary, obtain the best gas mileage, top operating efficiency, longer life, and fewer costly repairs by keeping the car in prime running condition. Good maintenance can add up to huge savings in mileage and dollars. A few simple maintenance measures include:[1]

- Follow the instructions listed in the owner's manual.
- Change the air filter every 3,000–4,000 miles. A dirty air filter can cut mileage by 10 per cent.
- Check the thermostat. When an engine runs too hot or too cold, gasoline is wasted.
- Adjust the choke. A slow or stuck choke can reduce the engine's efficiency by 30 per cent.
- Spark plugs should be replaced or cleaned and regapped every 5,000 miles. A single misfiring plug can cut mileage by 10 per cent.
- Under-inflated tires create tremendous friction with the road surface. This puts an extra burden on the engine and reduces gas mileage. Check the air pressure of tires every week. The owner's manual will give correct air pressure figures.
- Have the oil level checked after every other fill-up. Generally, change the oil filter with every oil change. Universal joints and wheel bearings should be kept greased; check every other oil change.
- Maintain a proper water-antifreeze mix in the radiator. Check the water level once a week. Before climatic conditions become extreme, add antifreeze (summer and winter) to prevent overheating or freezing of the car's cooling system.
- Check the transmission fluid level once a month while the engine is running.

These simple suggestions can help prevent major engine malfunctions from developing; however, professional work and servicing may sometimes be required. Try to use the same mechanic every time the car is serviced. Establishing a personal relationship with the mechanic and the service manager can improve the chances of good repair work. To find a reputable shop and certified mechanic, send for a copy of "Employers Directory," National Institute for Automotive Service Excellence, Suite 515, 1825 K Street, NW, Washington, DC 20006.

Before servicing a car check the warranty to see if the work is covered by the manufacturer. Describe in detail the problems to be corrected. Road test the car immediately after having repairs and return it to the garage if it is not performing well.

If one receives poor servicing or inadequate repairs and the repair shop does not recognize the complaints, contact the local Automotive Consumer Action Panel Office. "Autocaps" have been started across the country by the U. S. Office of Consumer Affairs. For the closest center, write to the National Automobile Dealers Association, 2000 K Street, NW, Washington, DC 20006. One may also appeal to the local Better Business Bureau or contact the nearest "Call for Action."

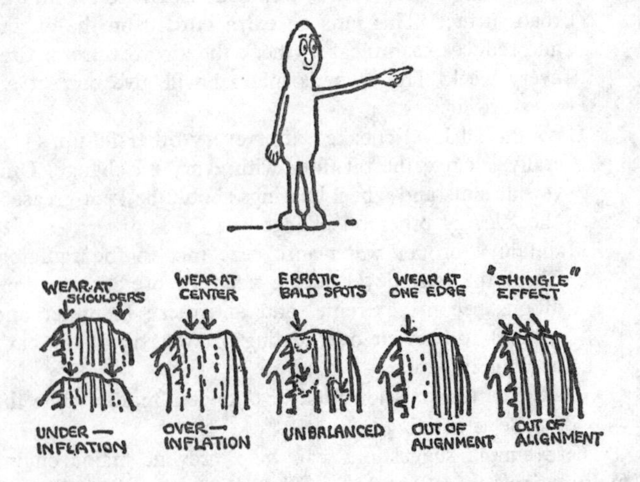

Note

1. "Car Care and Service," Consumer Publication, stock #2200-00090, General Services Administration, Superintendent of Documents, U.S.G.P.O., Washington, DC 20402, 1974.

Additional Sources

Center for Auto Safety
Dupont Circle Building
1346 Connecticut Ave., NW
Washington, DC 20036

National Call for Action
1785 Massachusetts Ave., NW
Washington, DC 20036

What to Do With Your Bad Car, by Ralph Nader, Lowell Dodge and Ralf Hotchkiss, Grossman Publishers, New York, NY, 1971.

"How Your Car Works," by Sam Dulty, *Times Minor Magazine,* New York, NY, 1974.

How to Keep Your Volkswagen Alive, by John Muir and Torg Gregg, Book People, Berkeley, CA 94710.

Popular Mechanics Complete Car Repair Manual, Popular Mechanics Books, New York, NY, 1974.

Chilton's Auto Repair Manual 1976, Chilton Book Co., Radnor, PA, 1975.

X COMMUNITY

> Now the company of those who believed were of one heart and soul, and no one said that any of the things which he possessed was his own, but they had everything in common.
>
> *Acts* 4:32

INTRODUCTION

Many forces have eroded local community: automobiles, television, suburban sprawl, and impersonal government. Public housing for low-income people destroys older "inefficient" neighborhoods, and replaces them with bland buildings devoid of life. Superhighways, shopping centers, and individual homesteads require much land and hasten the community's decline. People seem too busy to care about local education, housing, and land use. The community is suffering from impersonalism and human neglect.

For a strong community, decision-making power should rest in the hands of the local residents. Popular sovereignty is the cornerstone of this nation. In America's early history, local governments were independent and largely self-reliant. However, this spirit has been progressively lost, partly through centralization and urbanization. Retrieving this heritage should be a community priority. A new social and political consciousness must be raised wherein sharing, belonging, and caring for the needs of neighbors becomes important once again.

We must become involved in local civic, social, and political groups. Establish a group that is willing and able to answer local needs and bring change in areas such as municipal budgeting, social facilities, and health care. Gain neighborhood control by attending town meetings, running for office, and volunteering to research information. Bring the community back to life by participating in local affairs.

Additional Sources

Taking Charge, a Process for Simple Living: Personal and Social Change, American Friends Service Committee Bookstore, 2160 Lake St., San Francisco, CA 94121, 1975.

Small is Beautiful: Economics as if People Mattered, by E. F. Schumacher, Harper & Row, New York, NY, 1973.

The Limits to Growth, by Donella Meadows, Signet Books, New York, NY, 1972.

Moving Towards a New Society, by the Movement for a New Society: 4722 Baltimore Ave., Philadelphia, PA 19143, 1975.

Communities Magazine: Box 426, Louisa, VA 23093.

"Creative Simplicity Newsletter," from the Shakertown Pledge Group, West 44th St. & York Ave. South, Minneapolis, MN 55410.

Sharing Smaller Pies, by Tom Bender: 760 Vista Ave., SE, Salem, OR 97302.

1975 Alternative Celebrations Catalogue, from Alternatives: 1924 E. Third St., Bloomington, IN 47401.

Community Service
114 E. Whilema St.
Yellow Springs, OH 45387
(write for book list)

#86 LIVE COMMUNALLY

Two opposing trends concerning American households have developed in the last decade. One trend pertains to the growing number of small living units. Households in this country rose from 63,401,000 in 1970 to 69,859,000 in 1974, a 10 per cent increase, while the national population rose only 4 per cent. Smaller housing units are a greater drain on our energy supplies than any other consumer practice, since each unit requires individual heating and cooling facilities and appliances.

The second trend is the movement toward communal living. Some 2,000 communal groups were established between 1965 and 1970, and undoubtedly far more since then. With our cherished concept of private property, it is no wonder the commune seems threatening; for in a commune much is shared commonly. The largest misconception in the public mind is that the commune serves as an escape from reality. However, residents attest that instead of escaping social responsibility, they are moving toward a new and better society. Actually, an escapist would be unable to deal with the realities faced in communal living.

Communal living is as old as man—Iroquois long houses, Bedouin tent clusters, Tibetan monasteries, African bush villages, and early American settlements such as Jamestown or Plymouth Colony. These groups were building blocks of human civilization. Several moderate trends toward communal living can be found in condominiums and the rehabilitated sections of certain cities—Adams Morgan in Washington and Soho in New York City.

The contemporary commune represents many ideologies. One based on an ideological conviction, such as a religious community, will usually outlive a secular one held together merely out of convenience. Solidly based communes are interested in minimizing the essential needs of life through sharing with others and committing their actions to the betterment of society as a whole. Self-sufficiency, a traditional part of the commune idea, is not essential. The Walden Two experimental community finds that self-sufficiency is not even practical.[1] (They found the price of ready-to-eat chicken far below the price of chicken feed alone.) Though not self-sufficient, the community strives to be self-contained; it takes from the outside world what is needed and strives to minimize use of consumer products.

These groups encourage supportive endeavors such as farming, pottery, and weaving which provide essential capital and personal satisfaction. Work is accomplished, skills are developed, harmony is maintained, and a few quiet places are preserved for individual members. Human good is a collective endeavor; competition is de-emphasized.

> We are trying to be instruments of the cosmic forces working within the order of nature. We believe that earth, air, fire and water belong to everyone and can't be bought, sold, or owned. We are total revolutionaries; we are free men living with free creatures in a free universe.[2]

Community living saves resources: more persons per appliance, larger portions per cooking operation, less heating and cooling per capita, car pooling, etc. The ideal size of a commune depends upon the culture, resource availability and temperament of participants. What is an ideal living experience for one group may not be for another. Usually, when communities reach 50 or more residents, a structure becomes more visible and bureaucracy sets in. Too many meetings and procedures develop and life loses some of its personal touch. There are critical limits to size, but the ultimate success of the commune depends on the non-quantifiable spirit of the group.

MOVE CLOSER TO COMMUNAL LIVING

- Share housing and appliances with others.
- Encourage live-ins in exchange for services such as baby-sitting, yardwork, or farmwork.

- Convert unused rooms into apartments or rent bedrooms.
- Trade houses with out-of-towners for vacation housing.
- Bring in students for the summers or for particular events.
- Defend the rights of communal residents to live in the neighborhood.
- Visit a commune and see what they are like.
- Take the Shakertown Pledge (see p. 365).

Notes

1. *A Walden Two Experiment,* Kathleen Kinkade, William Morrow and Co., New York, NY, 1973, p. 100.
2. *Drop City,* by Peter Rabbit, Olympia Press, Inc., New York, NY, 1971.

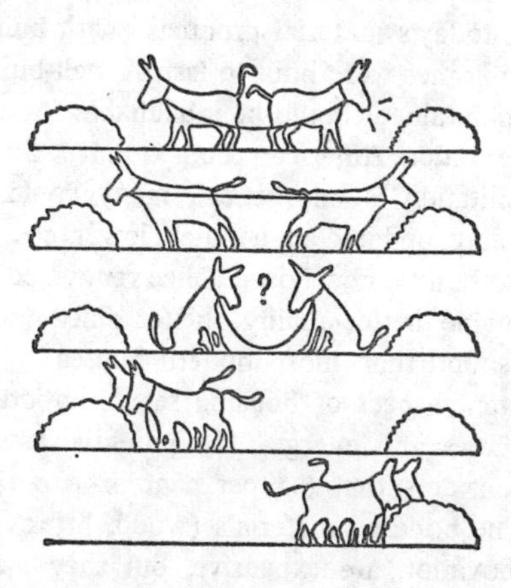

Additional Sources

The Art of Community, by Spencer H. Calhoun, Institute for Humane Studies, 1177 University Dr., Menlo Park, CA 94025, 1970.

The Family, Communes and Utopian Societies, edited by Sallie Teselle, Harper Torchbook, New York, NY.

Modern Utopia, by Alternatives: P.O. Drawer A, Diamond Heights Station, San Francisco, CA 94310.

The Joyful Community, by Benjamin Zablocki, Penguin Books, Baltimore, MD 21207.

Living Poor With Style, by Ernest Callenback, Bantam Books, New York, NY, 1972.

Frontier Living, The World Publishing Co., Cleveland, OH 44102.
Kibbutz: Venture in Utopia, by Melford E. Spiro, Schocken Books, Inc., New York, NY, 1975.
Communities, by Paul Goodman, Random House, Westminster, MD, 1947.

#87 REFURBISH OLD HOMES

Houses vary in age across the nation. Some are well maintained, but many older structures need extensive repair. Unlike many of today's material products (with built-in obsolescence), these homes were built to last. A well-built house that is properly maintained should be inhabitable for centuries. Although some older American houses have deteriorated to where rehabilitation is impractical, many could be restored and comfortably updated at a much lower cost than would be required to build a new house. Once renovated, older buildings can provide better-quality shelter since they are more structurally sound than most modern houses.

An important aspect of housing rehabilitation lies in conserving resources and energy. The potential savings is great when one considers that 6.5 per cent of America's housing is vacant.[1] The building materials (wood, brick, glass, metal) used for renovation are expensive, but they amount to far less resources and energy than the quantity required to construct a new dwelling.

Inner-city rehabilitation can help slow the expansion of suburban sprawl which resulted in people leaving the city core. Energywise, this means more efficient utilization of a city's resources. For example, an auto commuter working downtown could become a walker, biker, or mass-transit rider if his residence were close to his work. This would help reduce air pollution, noise, and congestion from auto travel (see Section IX).

If people returned to urban areas, the land used for suburban houses, stores, and roads could be left as open space. A reversal of current trends to move out of the city would encourage a new outlook on land use practices.

Housing renewal also has positive social benefits. Rebuilt neighborhoods can rekindle community pride and spirit, particularly in ghetto areas and depressed rural towns. Decaying sections of town, formerly considered to be unappealing, can be revitalized to stimulate business and activity. Property values would increase and provide a broader tax base for the city. The general appearance and character of sections of town would be vastly improved.

Community groups and religious organizations can cooperate with local redevelopment and renewal projects to provide housing for the poor. For example, one group in Washington, DC, Cornerstone Inc., is dedicated to renovating dilapidated housing for the poor and for volunteers who are serving the poor.

Housing rehabilitation can also help relieve the current employment crunch that is hitting the construction industry. Local contractors could renovate and improve existing buildings, putting people to work and helping to stimulate local community economies.

A house's potential for rehabilitation can be evaluated only by a systematic inspection, comparing cost of needed repairs with the value of the finished product. If the foundation is in good condition, and the roof, framing, floors, and walls are sound, the house is probably worth rehabilitating.

There are a number of "How To" books that describe all facets of interior and exterior home repair.

Refurbishing Hints

- Home repair can be accomplished simply and economically by anyone. Start with easy repairs before tackling more difficult ones.
- Become involved in a community project that benefits the poor.
- Promote urban and rural homesteading.
- Patronize a local contractor.
- Get at least three free estimates for each job.

- Have the house appraised by a professional house inspector.

Note

1. *The 1975 U.S. Factbook,* Bureau of the Census, Grosset & Dunlap, New York, NY, 1975, p. 698.

Additional Sources

Know-how: A Fix-it Book for the Clumsy but Pure of Heart, by Guy Alland, Miron Wakiw, and Tony Hess, Little, Brown and Co., Boston, MA, 1971.

The Complete Handyman's Do It Yourself Encyclopedia, by the editors of *Science and Mechanics,* Stuttman Company, Inc., New York, NY 10016.

House and Building Maintenance, James C. Woodin and Louis E. Hayes, McKnight Publishing Co., Bloomington, IL, 1969.

How to Buy and Remodel the Older House, Hubbard H. Cobb, Macmillan and Co., New York, NY, 1970.

How Things Work in Your House and What to do When They Don't, by the editors of Time-Life Books, Time-Life Books: New York, NY, 1975.

Wiring Simplified, H. P. Richter, Park Publishing Co., Minneapolis, MN, 1974.

Fundamentals of Carpentry, E. W. Sundberg and W. E. Durbahn, American Technical Society, 848 E. 58th St., Chicago, IL 60637, 1967.

Simple Home Repairs, U.S.D.A. Booklet, Superintendent of Documents, U.S.G.P.O., Washington, DC 20402.

Old Glory, America the Beautiful Fund, Warner Paperback Library, New York, NY, 1973.

#88 PROVIDE PURE DRINKING WATER

Most Americans consider their drinking water to be safe. However, recent studies indicate that the quality of public drinking water is rapidly declining. In 1969 a study of nearly 1,000 public water systems supplying 18 million users showed that almost 2.5 million people were drinking substandard water, and 360,000 were drinking water that was potentially dangerous.[1] Among smaller systems, one out of five delivered substandard water. Some experts feel that all surface water is polluted, and half of America drinks surface waters that must be treated to remove solids and kill bacteria. Other Americans drink underground (well and spring) water, but these sources are subject to poisons leaking from dumps and landfills, from water pumped underground to recover oil, and from chemical wastes pumped underground. It is difficult to find drinking water today free from chemical pollutants.[2]

The Environmental Protection Agency (E.P.A.) conducted a study of the lower Mississippi River above New Orleans, where a number of chemical plants and refineries are located. The study revealed heavy pollution, since the Mississippi, like many other rivers across the country, serves as a disposal site for industrial wastes and a source of drinking water. As a result, New Orleans, downstream from these industries, has heavily contaminated drinking water.

Increasingly, we are drinking waste water that has not been effectively treated. Almost sixty of America's public water systems draw on supplies where polluting industries are upstream from drinking-water intakes.[3]

Water is an excellent solvent, a property that makes it necessary for life. This property exacerbates our drinking-water pollution problems. In its journey to the faucet, water often comes in contact with toxic substances. For example, falling rain reacts with toxic gases and collects particulate matter from the air. Sulfur oxides from fossil fuels react with water vapor to form acid rain, which is highly corrosive and harmful to man, animals, and plants.

Rain water is absorbed into the substrata or collects in lakes, rivers, or streams that serve as drinking-water reservoirs. In both situations, water can be easily contaminated with toxic substances. As runoff from agricultural lands, water dissolves animal wastes, fertilizers, pesticides, and minerals. Industrial and municipal waste systems contribute additional contaminants.

THE HAZARDS

Toxic Metals

Toxic metals in drinking water result primarily from industrial waste disposal. For example, one plant in the lower Mississippi River discharges 3,700 pounds of lead in a single day.[4] Lead in drinking water is especially serious because it accumulates in one's body and can damage the kidneys, liver, brain, reproductive organs, and the central nervous system. Children, in particular, are affected by lead poisoning which can result in mental retardation. Other dangerous chemicals found in industrial wastes include arsenic, barium, cadmium, chromium, cyanide, fluoride, mercury, nitrate, selenium, and silver.

Nitrates

Agricultural communities often have high nitrate levels in their water systems caused by local fertilization. Nitrates in drinking water are extremely hazardous because they are converted in the body to nitrites which interfere with the blood's ability to transport oxygen; they are also carcinogenic agents. A link between nitrate intake and methemoglobinemia in infants has been firmly established. Since its first identification in 1945, more than 1,000 cases of infant methemoglobinemia have been reported throughout the world, including eighty-three deaths. Most cases were associated with high levels of nitrate in the water used for infant formulas.[5]

Chlorine

Recent studies have shown high concentrations of chloroform and carbon tetrachloride in the drinking water of Cincinnati and New Orleans. These toxic compounds are probably formed by the reaction of pollutants with the chlorine used for disinfecting the water. It is noteworthy that many countries use ozone for disinfecting instead of chlorine.

Chlorinated Hydrocarbons and Other Organic Chemicals

A less obvious but more insidious threat to human health arises from long-lived chemicals in chlorinated hydrocarbon pesticides, such as DDT. PCB's (polychlorinated biphenyls) are chemicals similar to DDT in their properties and are even more prevalent in the environment. Their high degree of stability and low flammability make them ideal for many industrial uses, but human health risks associated with them may be extremely serious.

According to an E.P.A. study, 66 organic chemicals have been found in fresh water.[6] Some 33 of these were tested for carcinogenicity with 12 causing cancer. In addition, many were found to be mutagenic (genetic defects) or teratogenic (birth defects).

Biological Hazards

Sewage contributes phosphates, nitrates, and organic matter to our water supplies. This can help breed pathogenic bacteria that can cause diseases such as dysentery, shigellosis, typhoid, gastroenteritis, and infectious hepatitis. The bacteria do not necessarily result entirely from sewage outflow, but they may indicate the presence of soil bacteria. These, too, may be pathogenic in organically polluted water.

The quality of our nation's drinking water is questionable. Our drinking water is becoming a transmitter of disease and chemical contaminants. Current standards and treatment processes are not adequate to provide pure drinking water, as long as the sources from which communities draw water receive increasing amounts of domestic wastes, agricultural runoff, and industrial effluent. The solution lies in reducing the sources of water pollution. To help fight the declining quality of drinking water, contact:

Citizens Drinking Water Coalition
1875 Connecticut Ave. NW, Suite 1013
Washington, DC 20009
(202) 462-0505

Environmental Defense Fund
1525 18th St. NW
Washington, DC 20036

Or write a letter to the local, state, and federal representatives.

CITIZEN ACTION FOR CLEAN WATER

- Get the facts about local water. Make sure it is being analyzed and monitored.
- Investigate local water-pollution control programs and support them if good; insist on better ones if they are weak.
- Help citizen groups working on pollution.
- Don't pollute by littering beaches, parks, or waterways.
- Clean up after picnics.
- Work to establish a program to test for all contaminants at the local water-treatment station.
- Build a cistern for drinking water (see Diagram).
- Where water supplies are suspect, boil before drinking it.

Notes

1. "What Is Happening to Our Drinking Water," by Eugenia Keller, *Chemistry,* February 1975, p. 16.
2. Testimony by Carl Clark at Environmental Protection Agency Hearings on Safe Drinking Water Act of 1974, September 5, 1975. *Federal Register,* Vol. 40, August 7, 1975, pp. 33, 224–33, 238.
3. Keller, op. cit., p. 18.
4. "Drinking Water," by Janice Crossland and Virginia Brodine, *Environment,* April 1973, p. 12.
5. "Nitrites, Nitrates and Methemoglobinemia," by Douglas H. D. Lee, *Environmental Research,* Vol. 3, pp. 484–511.
6. "Organic Chemical Pollution of Freshwater," Water Quality Criteria Data Book: Vol. I, E.P.A. Water Quality Office, Washington, DC.

Other References

"The Safe Drinking Water Act of 1974," E.P.A. Pamphlet, Superintendent of Documents, U.S.G.P.O., Washington, DC 20402, July 1975.

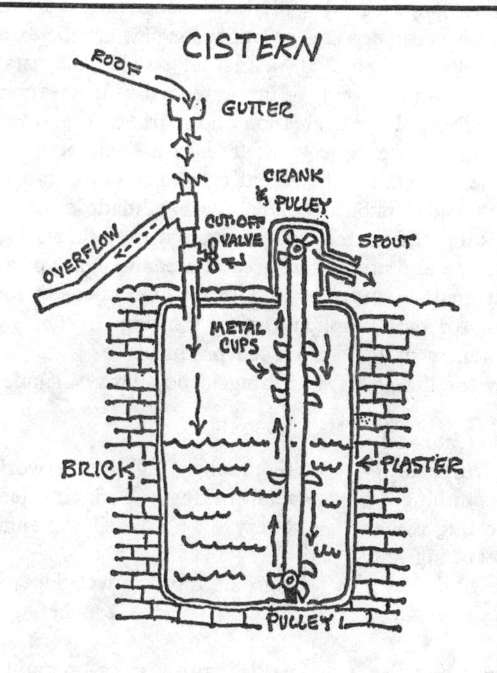
CISTERN
ROOF
GUTTER
CRANK
& PULLEY
CUT-OFF
VALVE
SPOUT
OVERFLOW
METAL
CUPS
BRICK
PLASTER
PULLEY

#89 JOIN A CRAFT CO-OPERATIVE

Craft co-operatives are springing up all over the country in response to the needs of individual craftsmen. Exchanging knowledge of craft techniques, preserving these techniques, creating a market for co-operative members, and sharing expenses for materials, tools, and rent can be accomplished more easily by an organized group than by an individual.

Co-operatives serve as alternatives to the regimentation and restrictions imposed by galleries or academic communities. Co-operatives can serve as an outlet for new artistic expression and creativity and provide a market outside the established galleries. All too often, a first-rate artist or craftperson is denied a channel for expression and marketing, and must discontinue his/her artistic work to earn a livelihood.

Another reason for the recent growth of craft co-operatives is the increased public demand for handmade crafts. People are resisting the abundance of standardized, mass-produced merchandise and are finding greater meaning and appreciation in crafts made by individuals. This reaction seems to represent nostalgia for past simplicity and a growing resistance against the blandness of mass-produced products.

A few of the crafts increasing in popularity include:

making musical instruments	lithography	leatherworking
blacksmithing	metalworking	making jewelry
decorating dishes	pottery	silkscreening
glass cutting and blowing	embroidering	weaving
making furniture	quilting	crocheting

Three principles usually distinguish a co-operative from other types of business:

- The co-op does not make money for itself—profits belong to co-operative members in proportion to patronage, resource contribution, labor, or some other basis for allocation.

- The structure is democratic—each member has one vote, regardless of labor or contribution.
- The return on earned money is limited—profit is not the main benefit or function of the co-operative.

The structure of each group will differ according to the nature and purpose (status, income, ambition) of the individuals involved. The craft co-op should question the basic assumptions of our consumer society and offer an alternative more responsive to artist and community.

Specific advantages include:

- Marketing assistance to the prospective seller. Each co-op needs a member who knows the market.
- Lower costs for supplies due to group purchasing.
- Technical assistance and training programs with skilled craftsmen as instructors.
- Quality control to establish product standards and inspire buyer confidence.
- Stimulating functional products and new designs by sharing techniques and knowledge.

If interested in organizing or joining a craft co-operative, the following groups may provide assistance:

Union of Third World Shoppes
428 East Berry St.
Fort Wayne, IN 46802

Office of Public Affairs
Economic Development Administration
Department of Commerce
Washington, DC 20230

National Collection of Fine Arts, Renwick Gallery; American Folklife Festival, Division of Performing Arts; and Museum of History and Technology, Cultural History Division, Smithsonian Institution, Washington, DC 20506

Federal Inter-Agency Craft Committee
National Endowment for the Arts
Washington, DC 20506

National groups that may give assistance:

American Crafts Council
29 West 53rd St.
New York, NY 10019

Associated Council of the Arts
1564 Broadway
New York, NY 10036

American Federation of Arts
41 East 65th St.
New York, NY 10021

Reference

The Cooperative Approach to Crafts, Program Aid No. 1001, Farmer Cooperative Service, U. S. Department of Agriculture, Washington, DC 20250.

Additional Sources

Publications from the Farmer Cooperative Service, U. S. Department of Agriculture, Washington, DC 20250.

The Book of Country Crafts, Randolph Wardell Johnston, Routledge and Keegan Paul, 9 Park St., Boston, MA 02108.

"The Potential of Handcrafts as a Viable Economic Force," from the Office of Public Affairs, Economic Development Administration, Washington, DC 20030.

Self-Help Program
Box M
21 South 12th St.
Akron, PA 17051 (send for a catalogue)

Washita Valley Crafts
Box 218
Anadarko, OK 73005 (send for a catalogue)

Maco Crafts (a self-help craft co-operative)
Box 1345
Franklin, NC 28734

Holston Mountain Co-op Crafts
279 East Main
Abingdon, VA 24210

International Program for Human Resource Development
PO Box 30216
Bethesda, MD 20014

Home Co-op
Route 1
Orland, ME 04472

Bolivian Handcrafts
21 Oak Dr.
Orinda, CA 94563

#90 CARE FOR THE ELDERLY AND ILL

For most people, identity as a human being draws strongly from one's home life. The home is a way of life, composed of neighbors, the church or synagogue, the grocery store . . . the familiar surroundings that encompass the world of an individual.

It would be nice to envision old age as a time of peace and comfort–a just reward for a life of hard work and productivity. But society cannot ensure this reward for many elderly; money, health care, living arrangements, transportation, and the lack of community activities are problems that plague the lives of the elderly.

There are twenty-two million Americans over sixty-five years old, and one quarter of these live alone. Although exact statistics for their living conditions do not exist, many experience malnutrition, isolation, and unhappiness. In a culture that places so much emphasis on youth, we ignore an important biological fact—that young people grow old every day. By the year 2000 the elderly will account for 25 per cent of the population. Our youth culture is aging.

Many elderly living in nursing homes and hospitals would be happier in homes with their families. The home environment fills emotional needs that can never be met by impersonal and expensive institutions. However, family members must be willing to make sacrifices.

The suitability of a residence for an at-home patient can be evaluated by the following check list:[1]

- Is the home quiet?
- Are there stairs going into the house? Inside the house?
- Is there room for a wheelchair or walker?
- Will the patient have a private room?
- Are there children in the house?
- What floor coverings are in the house?
- Is a telephone in the patient's room or nearby?
- Is an adult in the house to care for the patient during the day? At night?
- Is the room properly furnished to prevent boredom?
- Is a shower and/or bathtub available?

The so-called "disadvantages" of the daily family routine and of time-consuming responsibilities seldom outweigh the psychological, emotional, and economic advantages of in-home health care. A patient's ability to recuperate is often improved in one's home, rather than in a hospital. Hospitals across the country administer home-care programs providing equipment in the home, periodic visits by doctors, nurses, and other medical personnel, and special training for family members.

HINTS ON HOME CARE

- Talk to family members before the patient comes home. Co-operation is a necessity.
- Ask the doctor for advice on the level of care needed. Perhaps a combination of care facilities is required.

- Contact local home health care service organizations and ask for advice.
- Ask questions for the sake of both attendant and patient.
- Know the costs, billing procedures, and extra charges involved in having an attendant in service at home if one is needed.
- Check with an insurance company about coverage.
- Maintain expense records. Legitimate medical expenses are tax-deductible.
- Most importantly, prepare to sacrifice and change ordinary family routine. Taking care of an elderly person at home is not easy—it is a sign of love and requires something from all family members.

If one wants to care for the elderly at home but wants the advantages of nursing-home facilities, the day-resident program may be the answer. Day-care centers allow an individual to return home at night. A few objectives of the day-care center are:

- To provide an opportunity to participate in a variety of activities during the day and early evening.
- To provide a semi-protective environment while not giving the person a feeling of entrapment.
- Where appropriate, to reduce the need for permanent placement in a nursing home.
- To provide a balanced diet.
- To relieve tensions in family situations at the regular home.
- To maximize the use of facilities without having to expand one's home.

Usually, day care is not open to everyone. Most participants are referred by a social-service agency. In most cases, the participants are ambulatory and do not need nursing care and supervision. Recreation programs, occupational therapy, religious, social, and dietary services are the primary functions provided. Meals on Wheels is a food service organized locally across the country to provide an adequate diet for the elderly and incapacitated.[2]

For more information, write:

United States Commission on Aging
U. S. Department of Health, Education, and Welfare
Washington, DC 20203

Various state agencies on aging.

National Association of Home-Delivered Meals
2209 21st Pl., SE
Washington, DC 20020

Local services—senior centers and clubs, health and welfare councils, special divisions of local health and welfare departments, and district offices of the Social Security Administration.

Private organizations—the National Council on Aging, the American Association of Retired Persons, and the National Council of Senior Citizens.

Railroad Retirement Board
844 Rush St.
Chicago, IL 60611

Civil Service Commission, Bureau of Retirement
Insurance and Occupational Health
Washington, DC 20415

Veterans Administration, Veterans Benefits Office
2033 M St., NW
Washington, DC 20420

Help the elderly obtain Old-Age Assistance. See whether they are eligible for social security benefits. They may also be eligible for Medicare, a federal program that helps pay for health care for almost anyone over sixty-five.

- Part A is the hospital insurance. It provides basic protection against the costs of inpatient hospital care, post-hospital extended care, and post-hospital health care.

- Part B is the medical insurance that provides supplemental protection against the costs of physicians, medical supplies, home health care, outpatient hospital services, therapy, and other services.

Medicaid, a federal-state administered assistance program, is designed to help certain kinds of needy and low-income people. Among those who qualify are: the aged (sixty-five or over), the disabled, the blind. Other forms of assistance and relief are: food stamps, tax relief for senior citizens, and discounts for transportation and recreational facilities, such as the Golden Age Pass which allows free entrance to all National Parks.

Contact the local health department for information on public health clinics, mental health clinics, health screening programs, and health maintenance organizations.

Notes

1. *What You Should Know About Health Care,* by Alan E. Nourse, M.D. and Mrs. Marie Burbidge, National Director of Consumer Affairs, Homemakers, Home and Health Care Services, Subsidiary of the Upjohn Co., Kalamazoo, MI 49001.
2. National Association of Home-Delivered Meals (Mrs. R. Kaczmarck)
   2209 31st Pl., SE
   Washington, DC 20020

Other References

The Old Folks At Home, produced by Imre Harvath and Lucy Salenger, "60 Minutes," Vol. VII, Number 28, as broadcast over the CBS Television Network, with CBS News correspondents Morley Safer and Mike Wallace.

Additional Sources

Old Age: The Last Segregation, by Claire Townsend, Bantam Books, New York, NY, 1974.

The Coming of Age, by Simone de Beauvoir, Warner Paperbacks, New York, NY, 1973.

Nobody Ever Died of Old Age, by Sharon R. Curtin, G. K. Hall & Co., Boston, MA, 1973.

Tender Loving Greed: How the Incredibly Lucrative Nursing Home Industry Is Exploiting Old People and Defrauding Us All, by Mary Adelaide Mendelson, Alfred A. Knopf, Inc., New York, NY, 1974.

Older Americans: Special Handling Required, by Marjorie Bloomberg Tiven, National Council on the Aging: 1828 L St., NW, Washington, DC 20036, 1971 (free).

Gray Panthers
3700 Chestnut St.
Philadelphia, PA 19104

Self Help for the Elderly
3 Old Chinatown Ln.
San Francisco, CA 94108

Senior Citizens Coalition of CAP
2200 N. Lincoln
Chicago, IL 60614

National Council for Senior Citizens
1511 K St., NW
Washington, DC 20006

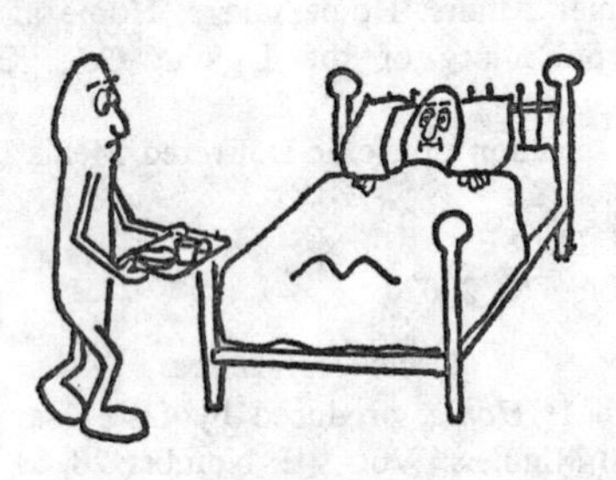

#91 HAVE SIMPLE FUNERALS

The manner in which the dead are buried in modern society is strikingly different from traditional culture. Today, the average citizen is far removed from direct participation in the burial. Death has been institutionalized. Hospitals take care of the dying and professional funeral directors deal with them from death to the grave. The outcome of depersonalized death is that many funeral directors and cemetery owners are profiteering at the expense of grieving families.

At one time in America's past, the family and community took care of the burials. Family and friends traveled great distances to attend wakes, comfort the grieving, make the coffin, wash and dress the body, and dig and fill the grave, usually free of charge. As one person compares old-time burials with today's:

> Th' family didn't have t' pay nothin'. They dug th' grave free of charge. Men went in together and dug th' grave. And you made th' burin' clothes, and you made th' box t' be buried in, and there wadn't no payin' goin' on. Th' preacher never charged for a funeral—for preachin' a funeral. They'll charge for funerals now, preachers will. They're not supposed to. See, most preachers is paid by salary, and that's one of his jobs. He ain't supposed t' charge y'. But many of 'em 'll take anything.[1]

In 1974 Americans spent $3.6 billion on burial arrangements. The casket alone costs far more than most people are able to pay. For example, a pine box costs $180; a pine casket, $300; and a steel casket, $800. After this point prices soar: a mahogany casket costs $2,500, and a copper one costs $3,000.

It is plain that the costs of a death to a family can be staggering, but what is more striking is the national land space problem. In the next decade, Americans will have to dispose of what amounts to the total populations of Ohio, Mississippi, Arkansas, North Carolina, and Alaska, creating an added problem for regional planners.

With increasing space limitations, rising costs, and environmental necessities, we should seek alternatives to the modern burial system. These are a few suggestions:

- Join a memorial society, which consists of a voluntary group of people who have joined together to obtain dignity, simplicity, and economy in funeral arrangements through advanced planning. A large family or close community would be ideal for this kind of arrangement.

Each society member would have a specific responsibility. Usually, a funeral director is appointed from the group who makes arrangements for the disposal of the body. Someone who has carpentry skills constructs a casket. A six-foot casket can be made inexpensively from wood bought from a lumber company. For example, a six-foot box can be made from plywood for approximately $5. From the Registrar of Vital Statistics, a death certificate can be obtained for $2.00. The memorial society can perform funerals with the same dignity as those of previous generations and cut costs drastically by avoiding expensive funeral homes.

- Cemetery plots tend to be expensive, but it is possible to have less expensive burials with as much dignity as a formal cemetery. Malcolm Wells, an architect, has designed a cemetery near Philadelphia that offers a solution to crowded cemeteries. Instead of being buried in a coffin, the body is wrapped in simple clothes and buried under three feet of earth in a park stocked with birds and animals.[2] Such a cemetery has no limits to the number of burials possible. This is the best answer to the land space problem. Other solutions are the "double depth burial" which buries more than one person in a plot. Another alternative would be to bury bodies vertically, which would provide more space in a given area of cemetery land. Finally, the method practiced in Europe could also be used in America. Instead of assigning one plot per person in perpetuity, the same graves are dug up after a certain time and used over and over again.
- Donate the body to a medical school. Anatomy classes need cadavers and state regulations require that all parts be properly buried after use. Heart, kidney, cornea, and other transplant operations are being performed. For further information on body donation write:

The National Society for Medical Research
111 Fourth St., SE
Rochester, MN 55901

The National Kidney Foundation
116 East 27th St.
New York, NY 10016

The Living Bank
PO Box 6725
Houston, TX 77025

Northern California Transplant Bank
751 South Bascom Ave.
San Jose, CA 95128

• Cremation is another method of dealing with land space limitations. The cost averages $150 and it is not illegal to cremate a body without a casket. In most states there are no restrictions on disposing of the ashes in any desired way. Cremation, however, does require a fuel energy expenditure.

Notes

1. *The Foxfire Book,* edited by Eliot Wigginton, Anchor Press/Doubleday: Garden City, NY, 1972, p. 306. (Reprinted by permission of Doubleday & Company, Inc.)
2. *Moneysworth: The Consumer Newsletter,* Vol. 4, #26, September 30, 1974.

Other References

"Ecology and Religion," the Reverend Dennis G. Kuby, B.D., *Minister Newsletter,* April 1974.

"Do It Yourself Burial for $50," by Anton Nelson, *The Updated Last Whole Earth Catalogue,* from the Whole Earth Truck Store, 558 Santa Cruz, Menlo Park, CA 94025.

Additional Sources

A Manual of Simple Burial, by Ernest Morgan, The Cello Press, Burnesville, NC 28714.

On Death and Dying, by Elizabeth Kubler-Ross, M.D., Macmillan Co., Riverside, NJ 08075.

Alternative Celebrations Catalogue, 3rd Ed., Alternatives, 1924 E. Third St., Bloomington, IN 47401, 1975.

The Price of Death, from the Seattle Regional Office of the Federal Trade Commission, Seattle, WA, 1975.

The Cost of Dying and What You Can Do About It, by Ray Paavo Arvio, Harper & Row, New York, NY, 1974.

#92 INFLUENCE LEGISLATION

Communication is the lifeblood of a community. The extent to which each community member or group can come together to discuss matters of concern is a measure of community vitality. Without some type of exchange there will be no self-awareness and no community growth. We may agree to these principles, but how many of us practice them?

Contacting one's elected representative and lobbying are important aspects of community involvement. Certain problems are handled more efficiently on the state or national level. A vocal and politically active local organization is the key to bringing change in the community, but any such effort requires effective leadership, intense personal commitments, and perseverance.

When one initiates legislative action remember that the chairperson is the key to communication; otherwise, all is chaos. Too often, a person is hooked on one position and uses the forum to browbeat the unconverted. The chairperson should identify this person early and harness such zealous enthusiasm. One way would be to have the person commit his/

her ideas on paper and have the speaker deliver the position at the appropriate time. Spontaneous vocal outbursts can upset community dynamics.

A. ORGANIZING FOR ACTION

- Identify the problem clearly; this requires thorough research of the parties involved in the dispute–who they are and what interests they represent.
- Be knowledgeable of all pertinent information. A well-read person will be able to respond to the opposition.
- Divide work according to abilities. Be sure that people with certain talents (speaking, writing, researching) are assigned those tasks.
- Identify allies and maintain close contact with them. Find local, state, and national groups with similar interests. Use their resources and experience and join with them.
- Keep the communication channels open to local newspapers and radio stations. Cultivate the mass media. They are best suited to spread the word.

B. STRATEGIES FOR ACTION

- Think of the pluses of an issue to be discussed. Critics often destroy ideas before they are allowed to develop. For an opening session, criticism must be controlled.
- After the meeting, reflect on what was said. Think of how much further a position can be developed, pro or con.
- List all negative aspects of the issue. Do not verbalize thoughts for their own sake, but for the community benefit.
- Repeat the reflection process after the meeting.
- At the next meeting, call for a consensus or majority vote on the issue. If major objections still persist, narrow the focus to be more specific.
- When a consensus is reached, the next step is to implement the strategy.

C. LOBBYING TECHNIQUES

- Find citizens who are knowledgeable, courteous, and convincing, and encourage them to lobby.
- Learn the elected representative's biography. See the *Congressional Directory* in the library.
- Know his or her background and political interests. Check the past voting record on similar and dissimilar issues. This will help determine where the legislator stands on public interest issues.
- Be familiar with past and current programs, associations, and committees of the target representative.
- Make an appointment with the representative.
- Know all counterarguments. It helps build self-confidence. Explain the position clearly and concisely. Provide the representative or aide with supplemental information such as reports, magazine articles, and scientific evidence.
- Follow-up is extremely important. Keep in contact with interested parties and watch the issue's progress.
- Keep in contact with the representative. Become a familiar face at the office.
- Know the legislative process. Be familiar with pressures on the legislator and what options are open to him/her. Sometimes, compromise is necessary.
- Contact related groups who are engaged in lobbying. Find out who they think are the best people to talk to and are the most responsive to community needs.
- Locate a source that can provide information on legislation (amendments, hearings, etc.) so that records can be kept up-to-date. Usually, this is obtainable through the elected representative or the state printing office.
- Be aware and keep track of various legislative alternatives dealing with particular issues.
- Follow the progress of other relevant bills.
- Watch for special-interest groups.

D. EVALUATING POLITICAL SUCCESS

- Assess the results of political action.
- Review the effectiveness of action periodically.

References

"Lobbying Tips," by Oren Heend, a sheet from the State and Local Organizing Project of the Consumer Federation of America: 1012 14th St., NW, Washington, DC 20005.

Additional Sources

The Consumer Activist's Handbook: A Guide for Citizen Leaders and Planners, by John Huenefeld, Beacon Press, Boston, MA, 1970.

FCNL (Friends Committee on National Legislation) "Washington Newsletter," 245 Second St., NE, Washington, DC 20002.

"Network Newsletter," 224 D St., SE, Washington, DC 20003.

The Organizer's Manual, O. M. Collective, Bantam Books, Inc., New York, NY.

The Environmental Law Handbook, by Norman J. Landau and Paul D. Rheingold, a Friends of the Earth/Ballantine Book, New York, NY, 1971.

Defending the Environment: A Strategy for Citizen Action, by Joseph L. Sax, Alfred A. Knopf, Inc., New York, NY, 1971.

Action for Change: A Student's Manual for Public Interest Organizing, by Ralph Nader and Donald Ross, Grossman Publishers, New York, NY, 1971.

Public Policy Reader, by Derek Shearer and Lee Webb, Institute for Policy Studies, Washington, DC 20009, 1975.

Neighborhood Power: The New Localism, by David Morris and Karl Hess, from Alternatives, 1924 E. Third St., Bloomington, IN 47401.

#93 CONSUMERS BEWARE

Living a simple lifestyle does not mean we cease being consumers. We must eat, clothe ourselves, and use materials to survive. We must purchase products from producers. We should be responsible consumers, however, defending our consumer rights and not buying in excess.

Consumer rights include:

- Safety—to be protected against the marketing of goods that are hazardous to health.
- Reliable information—to be protected against fraudulent, deceitful, or grossly misleading information, advertising, or labeling.
- Choice—to be assured access to a variety of products and services at competitive prices, and in those industries in which competition is not workable, to be assured of satisfactory quality and service at fair prices.
- A right to be heard—to be assured that consumer interests will receive full and sympathetic consideration in the formulation of government policy, and fair and expeditious treatment in its administrative tribunals.[1]

We also have a responsibility to other people and the environment when purchasing consumer products. As consumers we must:

- Reduce purchases of materials that needlessly drain our dwindling supplies of natural resources.
- Spread the word about harmful ingredients of certain products.
- Care for the environment by disposing of used products properly.
- Pressure marketers to follow fair pricing and labeling practices.
- Show concern about the occupational health of those who make consumer products.

The rise of consumer groups throughout the country reflects a growing outrage about the following abuses:

- Advertising pressures that create artificial demands by appealing to vanities, insecurities, and the desire for status.
- Pushing products that supposedly give instant remedies or improve one's social life.
- Planned obsolescence of products.
- Wanton depletion of natural resources.
- Major littering problems and disposal of packaging and discarded products.
- Creating fads and fashion-consciousness for faster turnover.
- Fraudulent, misleading, or deceptive advertising.
- Blatantly unsafe or unhealthy products.

CONSUMER HINTS

- Shop wisely. Examine the product carefully. Check the quantity and how it compares in quality and content with similar items.
- Resist sales pressure and buy only what is needed or can be used.
- Be sure the neighborhood store uses unit pricing (price per standard measure) and open dating of food items.
- Complain when necessary about faulty sales practices or products. Go directly to the source and talk to the manager. Keep track of checks, warranties, and sales slips. If the direct approach is unsuccessful go to the local Better Business Bureau.
- Become familiar with the law—the Truth-In-Lending Act, the Credit Liability Act, and the Warranty Act.
- Encourage good practices. Congratulate a merchant on quality merchandise and good service.
- Work with local and national consumer groups (see Additional Sources).
- Read materials that discuss the quality of consumer products; publications such as *Consumer Reports* or *Consumers' Research Magazine.*
- Avoid buying name brands at inflated prices when store brands are of equal quality.

- Be wary of come-on sales and be sure that what is advertised is in stock.
- Beware of price discounts on the product label, printed by the manufacturer. The regular price may be raised to compensate for this discount.

Take legitimate complaints to the following agencies:

- Suspected false advertising–Federal Trade Commission.
- Meat and poultry products–U. S. Department of Agriculture.
- Restaurant sanitation–local health authorities.
- Products made and sold exclusively within a state–local or state health departments or similar law enforcement agencies.
- Suspected illegal sale of narcotics and dangerous drugs such as stimulants, depressants, and hallucinogens–Bureau of Narcotics and Dangerous Drugs, U. S. Department of Justice.
- Contaminated food–U. S. Food and Drug Administration.
- Unsolicited products by mail–U. S. Postal Service.
- Accidental poisonings–poison control centers.
- Dispensing practices by pharmacists and inflated drug prices–state pharmacy boards.
- Air and water pollution and pesticides–Environmental Protection Agency.
- Dangerous products–Consumer Product Safety Commission.

Note

1. Message to Congress, President John F. Kennedy, March 15, 1962.

Additional Sources

"Redress of Consumer Grievances," Report of the National Institute for Consumer Justice, Consumer Services Administration, Washington, DC 20506.

Directory, State, City & City Government Consumer Offices, Department of Health, Education and Welfare, Office of Consumer Affairs, Washington, DC 20201.

Call for Action, 1785 Massachusetts Ave., NW, Washington, DC 20036.

Consumer Research Magazine, Consumer Research, Inc., Bowerstown Rd., Washington, NJ 07882.
Who Put the Con in Consumer? by David Sanford, Liveright Publishers, New York, NY 10036, 1972.
"Energy Management Digest: Energywise Buying Practices," *Commerce Today,* May 13, 1974, p. 13.
Consumer Reports, Consumer's Union, Mount Vernon, NY 10550.

Consumer Federation of America
1012 14th St., NW
Washington, DC 20005

Consumer Product Safety Commission
1750 K St., NW
Washington, DC 20207

National Consumer Center for Legal Services
1750 New York Ave., NW
Washington, DC 20006

National Consumers Congress
1346 Connecticut Ave., NW
Washington, DC 20036

National Consumers League
1785 Massachusetts Ave., NW
Washington, DC 20036

Public Citizens Visitors Center
1200 15th St., NW
Washington, DC 20005

#94 BATTLE UTILITIES

Our nation's utility companies have power over the sale of natural gas and electricity. State and federal regulatory commissions are often too complacent to act against price fixing by these utilities. With spiraling energy prices, citizen vigilance and pressure is needed to curb utility abuse.

Rate structures usually discriminate against the small user who tries to conserve on energy use. During and after the Arab oil embargo, several power companies requested higher rates because their revenues dropped. The main reason for lower revenues was energy conservation by consumers rather than the fuel shortage. The companies felt the need to raise prices because of the decreased demand. Higher prices due to conservation is grossly unfair.

There are fair alternatives to the commonly used declining block pricing structure (which reduces prices for increasing usage):

- The flat-rate structure–a standard rate per kilowatt hour is charged regardless of the level of consumption.
- Inverted-rate structure–takes the declining block and turns it upside down so that large users would be charged more than low-level users.
- Peak-load pricing–imposes additional costs when a customer uses more energy than normal during a peak load time.
- Life-line service–allows the elderly and those on fixed incomes who consume small amounts of energy (less than 400 kilowatt hours) to pay a nominal charge and not be subject to rate increases.

An industry practice that should be more closely examined is the use of tax-accounting gimmicks. The industry has frequently exploited tax loopholes, and the savings have rarely, if ever, been passed on to the consumer. Income tax payments by major utilities declined from 13 per cent of revenues in 1956 to 2.6 per cent of revenues in 1973. By the authority granted to many regulatory commissions, the electric utility company is allowed to keep two records, one for the Internal Revenue Service and one for the commission; the savings are usually passed on to the stockholders.

The 52 power companies that paid no federal income tax in 1974 charged their customers for non-existent taxes totaling $269 million and received tax credits totaling $217 million.

Advertising is another cost that has increased rates for customers. While energy sources are dwindling all over the world, it does not make sense for utilities to promote greater consumption of energy. A recent bill passed in Connecticut forbids utilities from charging customers for promotional, image-building, or political advertising.

HINTS FOR CONSERVING THE UTILITY DOLLAR

- Become informed; do not rely on utility publications as the only source of information.
- Develop a consumer-information program on the local utility. Arrange meetings with a utility representative.
- Become acquainted with the utility commissioners and staff. The companies know who their friends and enemies are–so should the citizens.
- Find out: who owns and works for the utility. Check interlocking directorate ties with other companies.
- Learn how the local rate structure works.
- Learn where the utilities spends its money. Find out how much is spent on advertising, lobbying, dues, and charity.
- Publicize worthwhile facts about the utilities.

A utility bill of rights should include:

- Fair-share rates–Large users should not get reduced rates; everyone should pay an equal amount for an equal amount of electricity used–or at least the small consumer should receive the break.
- Lifeline rates–To provide needed electricity at an affordable price. This will encourage people to use less electricity and provide relief for small-home users and small businesses.

- Peak pricing–This rate would charge large users more for electricity when it is in highest demand, and less when electricity is in lowest demand. This may help prevent brownouts.
- End the Fuel Adjustment Clause–This clause allows utility companies to raise rates automatically when coal and oil prices go up.
- Economical and safe power–Nuclear plants are not reliable and have serious health and safety problems.
- A clean environment and job protection–Strip mining has not been properly regulated, and thus vast stretches of our country have been denuded.
- A conservation program–To save money being wasted on unneeded new generators.
- Balanced growth–That provides jobs but does not destroy the environment.
- Renter's rights–Protect people whose electric bill is included in their rent; these people should have the advantages of Lifeline and Fair-share Rates.
- Fair and impartial hearings of grievances and complaints –For protection from overcharge and other abuses.
- End of arbitrary and unfair power shutoffs.
- Notification of changes in rates and other policies–in advance of these changes.

(Ideas from the East Tennessee Energy Group, Knoxville, TN.)

References

Taking Charge: A New Look at Public Power, Environmental Action Foundation, 724 Dupont Circle Bldg., Washington, DC 20036.

"D.C. Power Information Packet," People Organized to Win Equitable Rates, CSPI Publications, 1757 S St., NW, Washington, DC 20009.

Overcharge, by Senator Lee Metcalf and Vic Reinemer, David McKay Co., New York, NY 10017.

How to Challenge Your Local Electric Utility

"The Power Line" (a monthly newsletter on electricity utility issues)

A Citizen's Guide to the Fuel Adjustment Clause

Phantom Taxes in Your Electric Bill

All available from:

Environmental Action Foundation
724 Dupont Circle Building
Washington, DC 20036

Additional Sources

"People & Energy," CSPI Publications, 1757 S St., NW, Washington, DC 20009.

Citizens' Guide to Nuclear Power, Center for the Study of Responsive Law: 1908 Q St., NW, Washington, DC, 1975.

An Organizer's Notebook on Public Utilities and Energy, edited by Dan Leaky, Public Utilities/Energy Project, 410 College Ave., Ithaca, NY 14853.

Commercial Energy Waste: An Extravagance No One Can Afford, Massachusetts Public Interest Research Group, 233 N. Pleasant St., Amherst, MA 02116.

How California Won Lifeline, from the Movement for Economic Justice, 1609 Connecticut Ave., NW, Washington, DC 20009.

Toward Utility Rate Normalization (TURN)
2209 Van Ness
San Francisco, CA 94109

Electricity and Gas for the People
2463 Prince St.
Berkeley, CA 94705

Energy Policy Project
1776 Massachusetts Ave., NW
Washington, DC 20036

Save America's Vital Energy, Inc.
702 West Main St.
Belle Plaine, MN 56011

#95 DEMAND CORPORATE RESPONSIBILITY

Many people involved in the simple lifestyle movement have ignored the issue of corporate responsibility. But is it possible to live apart from the dominating influence of multinational corporations? United States foreign policy–from the Monroe Doctrine to the Marshall Plan and beyond–aided and encouraged America's corporations to penetrate and exploit markets in the Americas, Asia, Africa, and Europe. With government as a frequent accomplice, corporations have also imposed a domestic brand of colonialism on the American public.

Reinhold Niebuhr, author of *Moral Man and Immoral Society*, insists that power, rather than goodwill or reason, is necessary to overcome institutions. In a healthy democracy, power rests with the citizens. If corporations have been irresponsible, it is partly our fault. Our compliance is a luxury we can ill afford in an age of environmental pollution and resource depletion. However, corporate employees and the general public must be familiar with and watch over corporate practices.

Granting the present corporate system, we can work to improve it. Irresponsible corporate executives, for example, are sensitive to bad publicity–especially if they deserve it. Tactics to establish fair practices include:[1,2]

- General information–Learn about the local corporations. What do they do? Who are the directors? Are they interested in the social and physical well-being of the community? Do they placate people by superficial civic activities while glossing over major injustices? Pressure their public relations personnel to correct bad practices. Beware of conferences, gifts, and free lunches. Meetings between civic groups and corporate representatives should be formal and serious.

- Specific information–Research whether corporate practices have been open and ethical. Check for past occurrences of misconduct. Librarians can help find back issues of news stories and articles that contain important information. Talk with employees to obtain additional material.
- Demonstrations–If malpractice is discovered, picket or hold a vigil in front of an official's home.
- Boycotting–An economic boycott can be effective (such as the grape, lettuce, and Farah boycotts). It may be time consuming and require good organization, but it can work. Smaller companies are more vulnerable to this method than are the larger ones.
- Stockholder proposals–An avenue of individual action within the corporate structure is the shareholder proposal. If a person owns one share of company stock, he/she is entitled to attend the company's annual meeting, vote on all issues presented to stockholders, and raise issues to be voted on. This has become a popular way to pressure companies to change established attitudes on a number of policies and operations, particularly on those affecting the environment. Contact: Investor Responsibility Research Center, 1552 K St., NW, Suite 806, Washington, DC 20005.
- Counter recruitment–Use persuasion to discourage skilled and talented job seekers from joining an objectionable corporation.
- Legal action–Public-interest law firms can force governmental agencies to enforce corporate regulations. Legal actions include lawsuits, interventions in regulatory committee hearings, complaints, petitions, negotiations, and lobbying.

- Media–Counter advertisements can diminish the corporation's ability to create needs for new products, consolidate markets, and start commercial trends.
- Support local farms, craft shops, small businesses, and alternative organizations.
- Assist organizations that work on exposing corporate malpractices, especially in food and environmental issues.

Notes

1. *Corporate Action Guide,* by Corporate Action Project, 1500 Farragut St., NW, Washington, DC 20011, 1974.
2. *Ethics in the Corporate Policy Process: An Introduction,* by Charles S. McCoy, Center for Ethics and Social Policy, Graduate Theological Union, Berkeley, CA, 1975.

Additional Sources

Corporate Power in America, by Ralph Nader and Mark J. Green, Grossman, New York, NY, 1973.

Who Rules America? by William Domhoff, Prentice-Hall, Englewood Cliffs, NJ, 1967.

The Rich and the Super-Rich, by Ferdinand Lundberg, Bantam Books, New York, NY, 1968.

Global Reach: The Power of the Multinational Corporations, by Richard Barnet and Ronald E. Muller, Simon and Schuster, New York, NY, 1974.

How to Stop Corporate Polluters and Make Money Doing It, by William H. Brow, Bellerphon Book Co., New York, NY, 1972.

Open Books: How to Research a Corporation, from Urban Planning Aid, Inc., Cambridge, MA, 1974.

Guide to Corporations: A Social Perspective, by the Council on Economic Priorities, Shallow Press, New York, NY, 1974.

The Consumer and Corporate Accountability, by Ralph Nader, Harcourt, Brace, Jovanovich, New York, NY, 1973.

A Public Citizen's Action Manual, by Donald Ross, Grossman Pubs., Inc., New York, NY, 1973.

People/Profits: The Ethics of Investments, from the Council on Religion and International Affairs, New York, NY, 1972.

Corporate Information Center
475 Riverside Dr.
New York, NY 10027

Center for New Corporate Priorities
1516 Westwood Blvd., Suite 202
Los Angeles, CA 90024

Clearinghouse on Corporate Social Responsibility
277 Park Ave.
New York, NY 10017

Corporate Accountability Research Group
1832 M St., NW, Suite 101
Washington, DC 20036

Corporate Democracy, Inc.
1165 Park Ave.
New York, NY 10028

Investor Responsibility Research Center
1522 K St., NW
Washington, DC 20005

Justice and Peace Center
3900 North 3rd St.
Milwaukee, WI 53212

#96 START RECYCLING PROJECTS

Many people regard recycling as a responsible environmental practice. In one sense it is, for we should reuse items that are serviceable, such as second-hand clothes and furniture, and we should save valuable resources by recycling paper and metals. However, industry often supports recycling in an effort to shift corporate responsibility to the consumer. Valuable copper or aluminum should be recycled, but only for necessary purposes. Source reduction and resource conservation are more in line with simple living, rather than recycling products that should not have been made in the first place.

Recycling does save energy over manufacturing products from virgin materials. However, recycling is more than a community project. It demands a change in the economic structure so that collecting and reprocessing industries can compete with those offered depletion allowances and reduced shipping rates for using virgin materials. A successful recycling program must reach beyond the community and often requires state and national assistance.

General principles for proper recycling procedures:

- Reuse serviceable consumer items (clothes, furniture, appliances).
- Reuse trash and garbage when possible (container reuse and composting).
- Recycle metals, paper, and lubricating oil (if all waste oil were recycled as it was during World War II, the nation would save 1.1 billion gallons of oil each year).
- Discourage the production of non-recyclable items (bimetallic beverage containers).
- Apply citizen pressure for regional and national recycling programs.
- Emphasize resource conservation and curbing initial use of products.
- Never buy non-returnables.

TO ORGANIZE A RECYCLING PROGRAM

1. Collect a solid work staff: office help, contact people, publicity personnel, printing help, and volunteer workers.
2. Give the program a name, symbol, and slogan.
3. Find a site large enough to store up to 3 tons of scrap; have it easily accessible, protected from vandals, and identified.
4. Have someone oversee the site at a given time. Have additional helpers so that a few individuals do not end up bearing the brunt of the labor.
5. Determine collection spots: polling sites, shopping centers, school yards, vacant lots, etc.
6. Have containers at the collection sites. Use 4-ply bags, large cartons, metal or cardboard drums. Fifty-five-gallon drums can be obtained from service stations. Mark the cleaned drums with the slogan.
7. Set dates and hours for collections.
8. The cost of trucking materials to scrap dealers should be covered by the scrap value. In general, bi-metallic cans bring $10/ton, steel cans $20/ton, and aluminum $200/ton.
9. Encourage people to prepare materials before contributing. Cans should be rinsed and labels removed. Flatten aluminum cans, remove ends and flatten steel cans, thoroughly clean aluminum containers.
10. Keep weight records of cans collected (8 steel and 20 aluminum cans/lb.).
11. Set up guidelines for the potential contributors. Have a service to pick up for the aged and infirm.
12. For recycling glass, obtain information from glass manufacturers.
13. Newspapers should be tied securely in bundles approximately ten to twelve inches high. Magazines must be excluded. Fire insurance is especially important for an indoor newspaper collection site.

References

Environmental Action, Vol. 2, No. 23, April 17, 1971.

Environmental Action, Vol. 5, No. 2, May 25, 1974.

The Recycler's Handbook, A guide for Community Action, Booklet by The Can People, Edison, NY.

Energy Activity Guide, National Recreation and Park Association, 1601 N. Kent St., Arlington, VA, 1975.

Additional Sources

"Aluminum Can Recycling Centers," The Aluminum Association, 750 Third Ave., New York, NY 10017.

Recycling, by Thomas Fegely et al., Rodale Press, Emmaus, PA 18049, 1975.

Energy in Solid Waste, from the Citizens' Advisory Committee on Environmental Quality, 1700 Pennsylvania Ave., NW, Washington, DC 20006.

Recycling Handbook: A Guide to Running a Recycling Project, from Recycling Information Office, Oregon Department of Environmental Quality, 1234 S.W. Morrison St., Portland, OR 97205.

Resource Recovery Magazine, edited by Donald Bell, Waheman-Walworth, Darien, CT.

Solid Waste Information Retrieval System
Technical Information Staff
Office of Solid Waste Management U.S.E.P.A.
PO Box 2365
Rockville, MD 20852

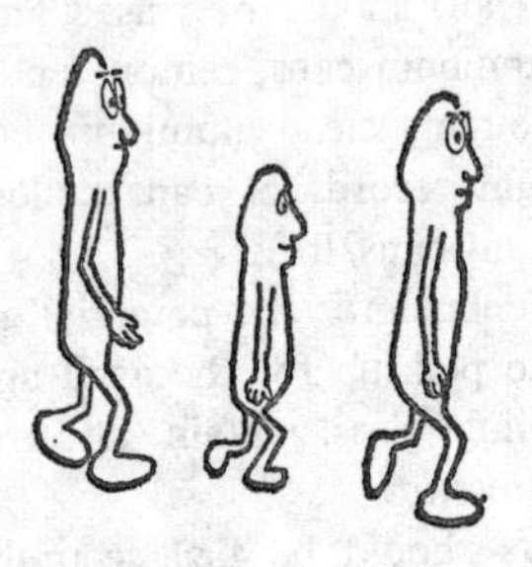

#97 FIGHT ENVIRONMENTAL POLLUTION

Environmental pollution affects our daily lives whether we realize it or not. Pollution is in the air we breathe, the water we drink, and all about us at work and in the home. No one in the world is entirely free from some form of pollution.

Air pollution corrodes buildings and other man-made structures, and endangers the health of the nation. In New York alone, deaths from emphysema have risen 500 per cent in the last ten years. The cases of asthma have skyrocketed in recent years; in Harlem, for example, 26 per cent of all patients in hospitals suffer from asthma as compared to 5 per cent in 1952.

Approximately 280 million tons of air pollutants are discharged into the air in the United States each year. This aerial waste contains nitrogen oxides, hydrocarbons, sulfur oxides, and carbon monoxide—all poisons to the human system. But health and environmental hazards are only part of the picture. The financial costs of polluted air are substantial and ultimately carried by the consumer. The U. S. Environmental Protection Agency (E.P.A.) estimates that a family in New York City spends an extra $850 for cleaning, household maintenance, and personal care due to environmental pollutants.

Water pollution is also a serious problem. The amount of pollution that annually enters our waterways has been estimated at over 400 million tons of suspended solids and 62 million tons of dissolved solids. Most of America's water-supply facilities were built over twenty years ago to clean the water of contaminants that were prevalent at that time (see entry #88). But contamination is only one type of water pollution.

Thermal pollution is another form that costs $13 billion a year. Industry, the largest user of water today, uses two thirds of the water available for cooling processes. Industrial cooling effluent averages 20° F. hotter than when it enters. This can have detrimental effects on the aquatic environment.

Noise pollution also takes a toll. An estimated 16 million Americans suffer some loss of hearing caused directly by noise. Noise pollution also causes other physical and psychological problems. At sound levels above 35–45 decibels, noise has adverse effects on a person in deep sleep. At levels above 85 decibels, stress reactions are likely to occur. Continual noise bombardment on any healthy individual has detrimental effects; loud penetrating noises can even affect the growth pattern of an unborn child.

The possible benefits from battling pollution are incalculable. The E.P.A. estimates that over a ten-year period, more than $260 billion could be saved through cleaning up pollution.

A community can mobilize against pollution through social, political, and legal actions. The possibilities include:

- Making professional, academic, and civic organizations more aware of environmental problems.
- Supporting political candidates who want to deal with environmental issues.
- Demanding corporate environmental responsibility at annual stockholders meetings.
- Volunteering to work with public interest and ecology groups. Retirees are of invaluable assistance to such groups.
- Joining citizen-action coalitions concerned with problems relating to highways, clean air, and land development.
- Picketing the homes of local polluters.
- Bringing suits and class actions in the name of a community that has been harmed by polluters. The Environmental Defense Fund, Sierra Club, Audubon Society, and the Natural Resources Defense Council are organizations that are willing to undertake environmental legal actions.

Check the following:

- What are local, state, and national pollution control regulations for air, water, rubbish, and dumping? How well are they enforced, and if they are not, why not? Who are the offenders?
- How does the local community dispose of trash, garbage,

and other solid waste? What is the quality of the drinking water? How effective are the waste-treatment plants, and where are they located?

- Do local industries, government, and commercial operations have pollution-control programs and equipment? If not, why not?

Other tips:

- A local group should use scientists, physicians, engineers, lawyers, and technicians from within the community to set environmental priorities and to help with community actions. Don't overlook local universities, colleges, and high schools as sources of information and expertise.
- Deal with a specific issue. For a new group it is best to concentrate on one particular problem or industry in the community. Suppress the temptation to tackle too many problems.
- Know what the state Environmental Protection Agency is doing. Watch them. The power of constant vigilance cannot be overestimated. Find allies on the staff who will provide information.

References

"The Campaign for Cleaner Air," Public Affairs Pamphlet No. 494, Superintendent of Documents, U.S.G.P.O., Washington, DC 20402, February 1974.

"Cleansing Our Waters," Public Affairs Pamphlet No. 497, Superintendent of Documents, U.S.G.P.O., Washington, DC 20402, 1974.

Noise Pollution, E.P.A. Pamphlets, Superintendent of Documents, U.S.G.P.O., Washington, DC 20402, 1974.

Man and His Endangered World, E.P.A. Pamphlets, Superintendent of Documents, U.S.G.P.O., Washington, DC 20402, 1974.

Don't Leave It All to the Experts, E.P.A. Pamphlets, Superintendent of Documents, U.S.G.P.O., Washington, DC 20402, 1974.

Additional Sources

Earth Tool Kit: A Field Manual for Citizen Activists, from Environmental Action, Pocket Books, New York, NY, 1971.

Defending the Environment, a Strategy for Citizen Action, by Joseph Sax, Alfred A. Knopf, Inc., New York, NY, 1971.

Environmental Demonstration Experiments and Projects for the

Secondary School, Thomas Brehman, Parker Pub. Co., Englewood Cliffs, NJ, 1973.

How to Kill a Golden State, William Bronson, Doubleday & Co., New York, NY, 1968.

A Clear View: Guide to Industrial Pollution Control, James Cannon, Inform, Inc., New York, 1975.

Work Is Dangerous to Your Health, by Jeanne M. Stellman and Susan Daum, Vintage Books, New York, NY, 1973.

The Environmental Impact Handbook, by Robert Burchell and David Listokin, Center for Urban Policy Research, New Brunswick, NH, 1975.

Environment Magazine, from the Scientists' Institute for Public Information: 438 N. Skinker Blvd., St. Louis, MO 63130.

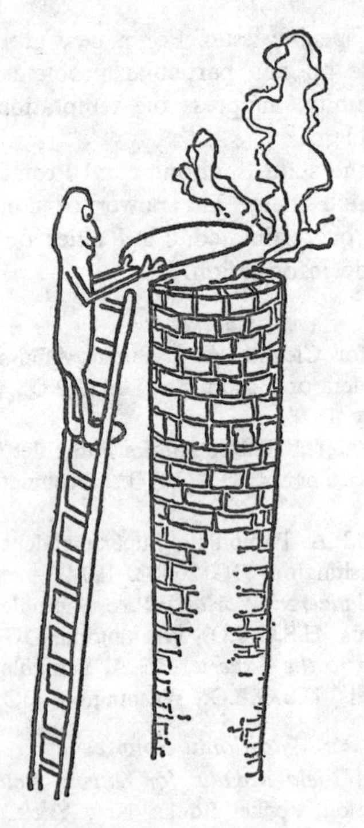

Natural Resources Defense Council
917 15th St., NW
Washington, DC 20005

Environmental Defense Fund
1525 18th St., NW
Washington, DC 20036

Sierra Club
1050 Mills Tower
San Francisco, CA 94104

National Audubon Society
950 Third Ave.
New York, NY 10022

Friends of the Earth
620 C St., SE
Washington, DC 20003

National Clean Air Coalition
1609 Connecticut Ave., NW
Washington, DC 20009

Citizens Against Noise
c/o Theodore Berland
2729 W. Lunt Ave.
Chicago, IL 60645

#98 IMPROVE LAND USE

Land use and abuse affects us in a number of ways. As a provider of food and energy, land is fundamental to the existence of man's economic structures. Open space is needed for peace and relaxation. Open habitat for wildlife is necessary to preserve the flora and fauna.

Land is a precious resource, but without proper zoning and land use regulations, it becomes a speculator's toy. Urban sprawl is a prime example of land misuse by developers.

Almost every issue–consumer, environmental, and energy –has its foundation in land ownership and use. High food prices result from monopolies that have virtually wiped out the individual farmer. Agribusiness giants own both fields and factories and, as a result, charge inflated prices for the food they produce. Energy and utility companies have ravaged the land through strip mining and have crisscrossed the landscape with powerlines. Highway builders have permanently removed thousands of acres of fertile land.

Citizens can approach this complex land-management issue through various avenues:

- Learn the facts. What are the local county-zoning regulations? Conservation easements? Community land trusts? Homesteading projects?
- Contact interested lawyers and find out what other communities are doing. Form a citizen's committee and compose model legislation.
- Expose unscrupulous developers by picketing their homes or offices. Bring citizen and legal pressure to bear on them.
- Find out who owns the land in the neighborhood. (An excellent model in West Virginia was written by Tom Miller of *The Herald Dispatch,* Huntington, WV 25701, "Who Owns West Virginia?")
- Check the local and state taxes and whether they favor withdrawing agricultural land from use.
- Fight bad development practices and be cautious of new industrialization schemes. Debate the issues in public long before the proposal land is purchased for construction. The earlier the better.
- Pressure large owners to allow gardening of unused property (see Section IV).
- Organizations to contact:

Sam Ely Community Land Trust
PO Box 116
Brunswick, ME 04011

Center for Rural Studies
1095 Market, RM 418
San Francisco, CA 94103

Southern Exposure
PO Box 230
Chapel Hill, NC 27514

Open Space Institute
145 E 52nd St.
New York, NY 10022

National Coalition for Land Reform
345 Franklin St.
San Francisco, CA 94102

Natural Resources Defense Council
917 15th St., NW
Washington, DC 20005

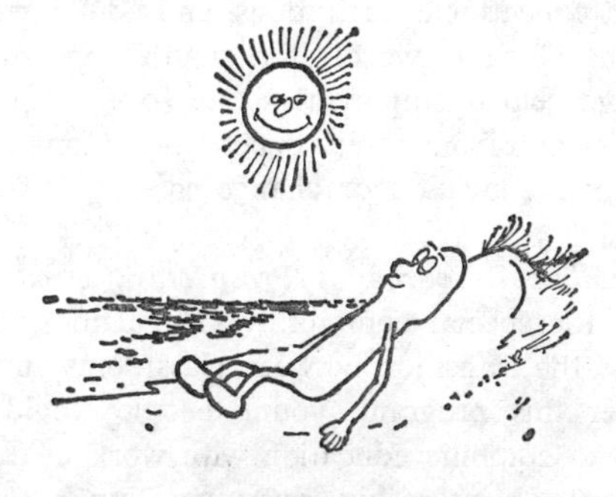

Additional Sources

Landscapes, by J. B. Jackson and Ervin H. Zube, University of Massachusetts Press, Amherst, MA 01002.

Protecting Open Space: A Guide to Selected Protection Techniques, by Elizabeth Kline, Society for the Protection of New Hampshire Forests, 5 South State St., Concord, NH 03301.

The Use of Land, by William Reilly, Thomas Crowell Co., New York, NY, 1973.

Design with Nature, by Ian L. McHarg, Doubleday & Co., Inc., Garden City, NY, 1971.

Land Use and the Environment, E.P.A. Reports, Superintendent of Documents, U.S.G.P.O., Washington, DC 20402.

Urbanization and Environment, by Thomas Detwyler and Melvin G. Marcus, Duxbury Press, Belmont, CA, 1972.

The Community Land Trust: A Guide to a New Model for Land Tenure in America, from the Center for Community Economic Development, 1878 Massachusetts Ave., Cambridge, MA 02140.
Guiding Growth: A Handbook for New Hampshire Townspeople, by Susan Redlich, Society for the Protection of New Hampshire Forests, 5 South State St., Concord, NH 03301.
The Quiet Revolution in Land Use Control, Superintendent of Documents, U.S.G.P.O., Washington, DC 20402.

#99 CREATE WORK AND STUDY OPPORTUNITIES

Many groups striving for change are isolated from one another. A common barrier exists between youths, who spend their time in educational institutions, and adults, who are absorbed in the working world. The youth population suffers from a 20 per cent unemployment rate (over 40 per cent are young blacks) and from career uncertainty. Many adults wish to further their education or change careers, but cannot do so very easily.

The Boundless Resource: A Prospectus for an Education Policy calls for special community "Education-Work Councils," to ease the transition between classrooms and the work place. Under this program, young people would have the opportunity to combine education with work in the community or to serve an internship. Every working or retired adult would also be entitled to an extra year of free education.

The Education-Work Councils would bring together people from schools, labor, and the community to explore ways for educational systems and industry to co-operate. In addition, the program would stimulate efforts to increase work satisfaction so that people would no longer feel trapped by their work conditions or educational deficiencies.

Among the recommendations of the program are:[1]

- Developing and administering programs providing young people with the opportunity for at least 500 hours of work per year.
- Instituting a comprehensive program of community in-

ternships and work apprenticeships that recognizes meaningful breaks in the formal education sequence.

- Creating a comprehensive outlook and career-information-reporting system by federal and local agencies to keep records on the training force to project future manpower needs.
- Removing state laws and regulations that bar adults from elementary and secondary public education.
- Establishing procedures ensuring that high school and college students receive at least five hours per year of career guidance and counseling.
- That all adults receive one year's deferred educational opportunity.
- That adults who have not yet received twelve years of formal schooling be entitled to as many years of free public education as they have missed, up to four years.
- That instruments for social planning be developed comparable to those commonly used for economic planning.

Full employment is a conservation measure. Our nation has many talents that are wasted when up to 8 per cent of our people are unemployed. It is possible to simultaneously create jobs, conserve energy and natural resources, and protect the environment. However, the American economy must shift its emphasis from undifferentiated industrial growth and technological development toward more useful employment and labor intensive methods of production. Some of the worthwhile government programs toward this end urged by the Council on Environmental Quality include:

- Accelerated construction of waste-water treatment systems, labor intensive, requiring light construction and unskilled workers.
- Highway repair: labor intensive, requiring light-construction activity.
- Public transit investment: involves repairing roadbeds for railroads and would employ unskilled workers.
- Strip-mine reclamation: increase employment in the economically depressed Appalachian region.
- Improving public lands: potential jobs—tree planting, range revegetation, erosion control, trail construction

and maintenance, wildlife habitat improvements, and litter control.

These programs should use labor from groups such as the unskilled and agricultural workers, who currently have high unemployment rates. These programs are able to start quickly, have the support of other national goals such as energy conservation, urban revitalization, and assistance to the poor, and produce long-term benefits.[2]

Other federal work programs under community control include:

- Economic Development Administration: to stimulate economically depressed areas with training programs and enterprises using labor intensive intermediate technology.
- Housing and Community Development: to rehabilitate blighted or deteriorated property for public works projects.
- Indian Financing Act: loans to Indian tribes and individuals for economic development on or near Indian reservations for educational purposes.
- Community Development Corporations: CDC is organized and controlled by local residents to develop the economy of a given community. CDC identifies and develops local skills and talents, owns and controls land and other resources, starts new businesses and industries, increases job opportunities, sponsors new community facilities and services, improves the physical environment.

For more information on these and other non-profit organizations, write:

National Congress for Community Economic Development
1126 16th St., NW
Washington, DC 20036

The Center for Community Economic Development
1878 Massachusetts Ave.
Cambridge, MA 02140

Notes

1. *The Boundless Resource: A Prospectus for an Education Work Policy,* by Willard Wirtz and the National Manpower Institute, New Republic Book Co., Inc., Washington, DC.
2. *Environmentalists for Full Employment,* 1785 Massachusetts Ave., NW, Washington, DC 20036, November 1975.

Additional Sources

Small is Beautiful: Economics as if People Mattered, by E. F. Schumacher, Harper & Row, New York, NY, 1973.

Growth and Implications for the Future, by Elizabeth and David Dodson Gray, Dinosaur Press, PO Box 666, Branford, CT 06405.

Working Loose, New Vocations Project, A.M.F.S.C.: 2160 Lake St., San Francisco, CA, 1971.

What Color is Your Parachute? by Richard Bolles, The Speed Press, Berkeley, CA, 1972.

How to Change the Schools, by Ellen Lurie, Random House, Westminster, MD, 1970.

How to Research the Power Structure of Your Secondary School System, by Bert Marian, David Rosen, and David Osborne, Study Commission on Undergraduate Education and Education of Teachers, Lincoln, NE, 1973.

Freedom of Information Clearinghouse: PO Box 19367, Washington, DC 20036.

The Workbook, edited by Katherine and Peter Montague, Southwest Research and Information Center, PO Box 4524, Albuquerque, NM 87106.

THE SHAKERTOWN PLEDGE

Recognizing that the earth and the fulness thereof is a gift from our gracious God, and that we are called to cherish, nurture, and provide loving stewardship for the earth's resources,

And recognizing that life itself is a gift, and a call to responsibility, joy, and celebration,

I make the following declarations:

1. I declare myself to be a world citizen.
2. I commit myself to lead an ecologically sound life.
3. I commit myself to lead a life of creative simplicity and to share my personal wealth with the world's poor.
4. I commit myself to join with others in reshaping institutions in order to bring about a more just global society in which each person has full access to the needed resources for their physical, emotional, intellectual, and spiritual growth.
5. I commit myself to occupational accountability, and in so doing I will seek to avoid the creation of products which cause harm to others.
6. I affirm the gift of my body, and commit myself to its proper nourishment and physical well-being.
7. I commit myself to examine continually my relations with others, and to attempt to relate honestly, morally, and lovingly to those around me.

8. I commit myself to personal renewal through prayer, meditation, and study.

9. I commit myself to responsible participation in a community of faith.

For background materials write to:

Shakertown Pledge Group/Simple Living Network
West 44th and York Ave. South
Minneapolis, MN 55410

INDEX

S19